POSTPHENOMENOLOGY

Northwestern University
Studies in Phenomenology
and
Existential Philosophy

POSTPHENOMENOLOGY

Essays in the Postmodern Context

Don Ihde

Northwestern University Press
Evanston, Illinois

Northwestern University Press
Evanston, Illinois 60208-4210

Paperback published 1995 by Northwestern University Press.

Library of Congress Cataloging-in-Publication Data

Ihde, Don, 1914–
Postphenomenology: essays in the postmodern context/Don Ihde.
p. cm.—(Northwestern University studies in phenomenology and existential philosophy)
Includes bibliographical references.
ISBN 0-8101-1098-9 (cloth)
ISBN 0-8101-1275-2 (paper)
1. Philosophy, Modern—20th century. 2. Postmodernism.
3. Phenomenology. 4. Technology. I. Title. II. Series:
Northwestern University studies in phenomenology & existential philosophy.
B804.I37 1993
142'.7—dc20 93-33183
CIP

The paper used in this publication meets the minimum requirements of the American National Standard for Information Sciences—Permanence of Paper for Printed Library Materials, ANSI Z39.48-1984.

For all the Humanities Institute group
who helped me more appreciate
postmodernism

Contents

Acknowledgments

The following chapters have been previously published as indicated and all appear here with permission of the publishers as listed:

Chapter 1 appeared originally as "Technology in the Renaissance," in *Interpreting the Italian Renaissance: Literary Perspectives,* ed. Antonio Toscano, Filibrary, no. 1, 1991 (Monographic supplements to the *Forum Italicum*).

Chapter 2 appeared originally as "New Technologies/Old Cultures," in *New Worlds, New Technologies, New Issues,* ed. Stephen H. Cutcliffe, Steven L. Goldman, Manuel Medina, and Jose Sanmartin (Lehigh University Press, 1992).

Chapter 3 appeared as "Technology as Cultural Instrument," in *Phenomenology and Indian Philosophy,* ed. D. P. Chattopadhyaya, Lester Embree, and Jitendranath Mohanty, published by the Indian Council of Philosophical Research (New Delhi: Motilal Banarsidass Publishers, 1992).

Chapter 4, "Image Technologies and Traditional Culture," appeared in *Inquiry,* Symposium on Technology and Human Values, vol. 35, nos. 3/4, published by Scandinavian University Press, Oslo, Norway.

Chapter 5 is forthcoming in the *Proceedings of the German Phenomenological Society* (Suhrkamp).

Chapter 7 originally appeared as "Text and the New Hermeneutics," in *Paul Ricoeur: Narrative and Interpretation,* ed. David Wood (London: Routledge, 1991).

While all the other chapters were given as lectures at different occasions and places, their appearance in publication is original with this volume.

Introduction: Postphenomenology

The title was inevitable. Today we live amidst the "posts." It is a postindustrial era, a postnuclear period, and there is postfeminism, postanalytic philosophy, and above all, postmodernism—so why not postphenomenology? All these "posts" are, perhaps, something like a technological society's substitute for previous metaphorical forests, within which one could get lost. What all the postmodern captures is the sense of transition, of a proliferating pluralism, and—for the nostalgic—a "loss of the centers" or "foundations."

For my part, I do not experience this nostalgia and do not bemoan the loss of foundationalism. Indeed, I have previously called the style of phenomenology I have practiced a "nonfoundational" phenomenology.[1] Postphenomenology is just another way of characterizing it as different from, but owing to its ancestry.

Husserl, Merleau-Ponty, Heidegger, and Ricoeur, all play an important background here, even if "transcendental egos," "subjectivity," and even "Dasein" are missing. This is not simply because there are newer and now trendier arrivals in the rage for Derrida and deconstruction or Foucault and power/knowledge. What I have sometimes called "generic continental philosophy" has not changed its habits. It still is dominantly a scholarly activity which finds its satisfactions in citing only European giants, interpreting and criticizing texts, and reconstructing intellectual histories. It's only that the newcomers now tend to replace the godfathers in the frequency of papers and publications.

The milieu, however, is much different now than this mere shift of "reverence texts" indicates. It is a milieu in which the old lines in the earlier battles between "continentals" and "analysts" have been softened. It is a milieu in which boundaries between disciplines—particularly within the humanities—have blurred due to the strong interests in humanities subject matters which have again recaptured much of the interest and imagination of philosophy graduate students and junior faculty. Indeed, I would claim that the previous focuses in language, logic, and science in hegemonic analytic philosophy, now finds a strong counterweight in interests in comparative literature and the literary and cultural studies which now are proliferating in North American universities. Some even say that comparative literature has replaced philosophy as the discipline which interogates human existence more primarily—but I find that the case only in those locations where philosophy itself has not recovered from its recent past. What I find, instead, is that philosophy is now more widely read and sometimes is captured more in the literary fields than in some academic departments.

It is a milieu in which "theory" abounds in a plurality of forms. These include the older neo-Marxian strands, a revival of neo-Freudianism (including Lacan and other revisionists), continuing phenomenological-hermeneutic strands, critical theory, feminisms, and ethnic cultural studies. No one has a hegenomy. Some are complementary, but others are in contest, and a broader front has been entered against older, more traditional forms of criticism and interpretation. Here, again, is our forest of "posts."

Where does what I am calling postphenomenology fit within this plurality of posts? To answer that I first turn to my own trajectory within North American philosophy. I was never, from the beginning, a genuine "transcendental" philosopner. Without knowing it explicitly, I was much closer to a very American set of traditions—Charles Sherover has sometimes characterized me as a "Heideggerian" Dewey, and Webster Hood saw this same connection. But only recently did I see how much that has always been the case.

The two philosophers who helped me realize this were Richard Rorty and Larry Hickman. Rorty—as many of us claimed when *Philosophy and The Mirror of Nature* came out—was nearly a decade and a half behind those of us who had agreed with the Europeans who had declared Cartesianism and even Kantianism to be "dead." There was a kind of nonfoundationalism already present in the hermeneutic and existential versions of phenomenology. But Dewey had claimed a kind of nonfoundationalism even earlier and Rorty picked up on that and its consequences in "late" analytic philosophy. His was an audience still mostly closed to Euro-Americans. And what Larry Hickman did with his *Dewey's Pragmatic Technology* was to show that Dewey actually preceded or at least paralleled what Heidegger had also done with respect to the primacy of praxis and even of the role of technology vis-à-vis science. Dewey's critique of the metaphysical tradition in many respects parallels Heidegger's, to which I was more attuned as a graduate student in the sixties.

What Dewey did not have, and what plays a very distinctive role in postphenomenology as I am defining it, was anything like the sophisticated theory of perception and embodiment which post-Husserlian philosophy developed (here I would note Merleau-Ponty, but also Heidegger and Ricoeur). Taking epistemology into a "motor intentionality" (Merleau-Ponty) or phenomenology "into the bloodstream" (Ricoeur) was to give me a deep and long lasting interest in the roles of perception and embodiment in our experience of our "world." My own contributions to that more precisely phenomenological task were first systematically developed in *Listening and Voice* (1976) and *Experimental Phenomenology* (1977), which respectively developed auditory and visual phenomenologies.

Yet, even before the printed versions of either of those books occurred, I became interested in the "extension" of perceptual and bodily intentionality into the smaller and larger "worlds" which were revealed through science and its instruments. Here was a perceptual technology, which became, from the early seventies on, the primary interest that has marked much of my career. And from 1979, with *Technics and Praxis,* most of my writing has had technology (and science) as a theme.

The claim of technoscience—as it is now increasingly called—put in phenomenological terms, is that it reveals a world which, perceptually identified, is both a microworld and a macroworld which could not be experienced except through the mediations of instruments. It was with that process that I became fascinated—but I also retained a kind of "instrumental realism" regarding the world-referentiality of this new mode of human perceptual embodiment.

This trajectory was already partially postmodern in that the modern or "Cartesian" subject was replaced with an existential "lived body," a kind of organism/environment model of interpretation which reembodied the human within a world or environment. The entire apparatus of "sensation," "sense data," even Husserlian "hyletic data," but also inferred material being, the "res extensa," disappears and is replaced by a more Merleau-Pontian "fleshly" interactive relativity. The "body" remains one strong thematic emphasis in the newer "theory" approaches, particularly within contemporary feminist discourse.

But within Euro-American philosophy, at least since the seventies, a countercurrent also began to gain strength. Its near relations included structuralism (which both Merleau-Ponty and Ricoeur also addressed) and its subsequent transformations into the poststructuralisms, including deconstruction. One of the claims of this movement was that the "subject" was decentered or displaced, first by virtue of a focus upon language as a system of signifier/signified, thus displacing the speaker/speech from a presumed privilege, and secondly in the move to see the "subject" itself as a kind of literary/historical construction.

In the process of establishing its differences from the phenomenological-hermeneutic traditions, proponents—including Derrida—often wildly exaggerated a presumed "absence" of such elements from phenomenology. As a result a virtual mini-industry of scholarly debates continue in conference circles.

What did emerge, however, as a shift in focus which began to occupy both the poststructuralist and hermeneutic traditions, was the elevation of the product of now displaced subjectivity or activity, the material residue, as it were, in the text.

This background is part of the reason why the essays in this collection display as their unifying theme a concern that arises with respect to one of the now major trends within Euro-American philosophy—its textism. Within the plurality of trends already identified, there is a whole spectrum of interests which, on the one hand, revolve around the fashionable metaphor of the text. And on the other hand, what remains of phenomenology is the practice of "reading" which is the activity related to texts.

Derrida is perhaps the most radical with respect to textism. His shift from the presumed primacy of speech, to writing and inscription, the trace, and later to textual marginalia, projects a kind of world-text refracted into its multiple fragments and possibilities. But Ricoeur, more soberly, also speaks of a world of the text. And Gadamer, and the later Heidegger, and in a degree, Foucault with his versions of discourse-praxis, help elevate "everything" to the level of "readings," "texts," and the like.

This emphasis is not without merit. In recent papers I have begun to argue that science itself shows many literary-critical traits and that philosophy of science should be seen as relating to science the way literary criticism relates to literature. In the philosophy of science, broadly construed, it has been Bruno Latour who has taken up this variant of the interpretation of science. In his groundbreaking *Science in Action* he even interprets laboratory experiments as "inscription making devices." But, this textism also worries me—and the more I take it seriously, the more I worry about it.

I always took phenomenology to be "realist" in aim. And although its referentiality was never causal, or linear, or direct as it was in the simpler analytic theories of reference, its aim was to describe the structural relativities between the knower and the world. And, in the worlds far beyond "eyeball" worlds revealed in science, I have never left my sense of wonder and awe. Science reveals, and although as with all revealing there is also concealing, our scientifically constituted world is in many ways richer than its prescientific predecessors.

In all this Foucault was also crucial. His histories of perception were highly suggestive. One can say that perception was, in a sense, "invented" in the modern era—perception taken in a certain way, is contemporaneous with science. But if a birth of perception can be so associated with the birthpangs of the modern—and can thus be taken to be postmodernly transcended—textism by the same token could be taken as a kind of throwback to a premodern tendency. The "world as book" is, after all, older than the perceived worlds of science—it is the premodern medieval world, or, in Derrida's case, possibly the world of the Talmud and its textual modeling praxis which makes the association. It is as if God "wrote" the world, but this is to anachronistically retroproject another literary activity back into creation.

Texts can and often do presume to be referential. But the other aspect of the postmodern which complicates the issue is the simultaneous rise of forms of "social constructionism." Philosophy of science itself has been implicated here. Ever since Kuhn's *The Structure of Scientific Revolutions* (1962) the philosophy of science has begun to proliferate into its own form of postmodern perspectivalism. When the social constructionists began to interpret the "products" of science as socially constructed products, the whole game was opened to a parallel to what had also happened in the humanities disciplines.

I had become used to the blend of textism/social constructionism amongst the literati, and to a certain extent excused it as the elevation of one's own activity to transcendent status—to have reified the literary, fictive act to transcendental position. But when it also began to appear within the precincts of science my worries deepened.

To this point I have been acknowledging some of the intellectual ancestry to this collection of essays. The essays themselves were written between 1985 and 1991, the first five of those years while I served as Dean of Humanities and Fine Arts at Stony Brook, and then following into a research year spent at the University of Sydney, Australia.

It was while serving as a dean that both the wider implications of postmodernism, as well as a more intimate acquaintance with the literary disciplines, began to make an impact upon my thinking. In their original contexts, the essays address two different audiences, one more clearly the audience within which a discourse concerning Euro-American or continental philosophy takes place, and the other a much wider, interdisciplinary audience. On the other hand, all the essays are marked by two recurring themes: one is the intimate connection between technology and culture, and the other might be called a perceptual-bodily referentiality.

I have therefore grouped the essays according to whether the first or the second theme dominates. But throughout both, the essays also respond either directly or indirectly to the postmodern context. I conclude in the epilogue with a brief reflection on the wider state of humanities as we approach the twenty-first century.

The essays in which technological issues dominate are, in effect, a resituating of the postmodern. Modernity, so differently defined according to discipline and perspective, usually is at least strongly differentiated from the premodern, which is often associated with the medieval. Thus while "modern science" is usually identified with the beginning of the seventeenth century (Galileo, circa 1600), and modern art much later, both are easily contrasted with the premodern and pre-Renaissance.

Similarly, the proliferation of definitions of the postmodern do not often mesh, yet what is common to the sense of the postmodern is that, somehow, we are aware that an era is ending, modernity. My take on this moment of transition and change is to "read" the era in terms of its omnipresent technological texture.

Part 1, "Postmodernity and Technoculture," begins with a reflection on the birthtime of modernity in the figures of da Vinci and Columbus, both as individuals caught up in the death of medievalism and the opening to "new worlds." But as the essays follow each other, there is also a turn to what I claim to be a unique form of contemporary crossculturalism in the phenomenon of pluriculture, which poses distinct challenges to much that we have taken for granted, particularly our "Eurocentrism."

Part 2, "Practicing Postphenomenology," emphasizes more particularly a certain theory of praxis and perception, often in the context of the contemporary Euro-American philosophical discussions now dominant in those circles. These philosophers, particularly Derrida and Foucault, and then Ricoeur and the later Heidegger who helped situate the former, are integrated here. My theme is directly aimed at the role of perception as it relates to contemporary continental textism. To state a thesis most bluntly, I show my worries to be less about the loss of subjects or authors, than I do about not being bodies or perceivers.

The same theme reemerges in different contexts. Part of postmodern pluralism consists in the proliferation of new perspectives. And perception can be both "microperceptual," as in our sensory bodies, and "macroperceptual," as in the cultural perceptions of our sedimented fields of lifeworld acquisition. But if and insofar as perception is invented, that invention continues to relate to the bodily polymorphy which gets displayed in what I call uniquely postphenomenology, a nonfoundational and nontranscentdental phenomenology which makes variational theory its most important methodological strategy.

Perhaps the thinkers who have most persevered the sense of bodily perception are today the feminists, who sometimes quite explicitly extend phenomenology into gender issues (as with Iris Young), or who even without direct connections to this tradition have rediscovered the need to see embodiment as an important aspect of all knowledge gaining and constructing activities.

Here we return to the vicinity of the current battles between the "new scholarships" and traditionalists. With the postmodern context there has also been a second lowering of transcendentality in what can be called the emergence of a kind of refractive perspectivalism. If feminists have not forgotten the body, and if the new scholarships have their sensitivity to other cultures, then the issue also begins to touch our current questions about a "liberal" curriculum. Both science-as-institution and our liberal arts must now face the same perspectival deformation: on the one hand, all positions become perspectival, but on the other there is a wider impact, a kind of pervasive perspective which makes us aware that so much of what we have taken for granted has been dominated by a narrower set of perspectives, those so clearly dominated by the implicit histories of "white, male, elitist" texts and values. These concerns will not go away by any reforms or returns to days and times now past.

Finally, postphenomenology may be seen to share with the other "posts" a deep sense of contingency and fallibility. It is perspectival in both a trivial and a significant sense. To acknowledge perspectivalism has always been a phenomenological conviction—but that conviction also places perspectives as variations within a variational framework. And it is this variational theory which I have taken to be the very core of phenomenology and which continues to operate within postphenomenology sufficiently recognizably to warrant the suffix on the compound word.

Although somewhat more than half the essays published here have appeared or are appearing elsewhere, most have not appeared in North American publications. The acknowledgments list these. I want to specifically acknowledge here the support of my own home university, the State University of New York at Stony Brook, for its sabbatical and research support, and especially thank the Australians who made me a Scholar in the Department of General Philosophy at the University of Sydney. I am also grateful for the warm welcome of those persons, particularly Denise Russell and Paul Crittenden. My fellow travelers, Linda and Mark, are never far from this process, too, and I thank them for making this often peripatetic life possible with a traveling "home."

PART 1

POSTMODERNITY AND TECHNOCULTURE

One of the dimensions of the postmodern context is the hyperawareness of "invention." To invent is to socially *construct,* often with the implication that anything *constructed* may equivalently be *deconstructed.* Foucault already claimed in *The Order of Things* that "man" is precisely such an invention: "Strangely enough, man . . . is probably no more than a kind of rift in the order of things. . . . man is only a recent invention, a figure not yet two centuries old . . . [and] he will disappear again as soon as . . . knowledge has discovered a new form."[1]

Similarly, according to postmodernists, *perception,* too, is a socially constructed, historical invention. I begin in this first part with what necessarily is a kind of anachronistic return to just such a "birthtime" of *modern* perception. Both da Vinci and Columbus, who figure in the first two essays, stand at the historical juncture that marks the close of the medieval and the birth of the modern. Both remain, even in what was subsequently thought to be much of their innovation, within a certain medieval imagination.

Leonardo's impact, I claim, is the "invention" of a certain way of *seeing.* In many ways he is the harbinger of the engineering style of design and seeing which today even dominates computer graphics. But in another, he was extending precisely the high medieval imaginations of figures like Roger Bacon, Bosch, Bruno and others who imagined strange creatures, worlds, and machines. Similarly, Columbus—who today is a radically different Columbus from the one celebrated in the quatercentennial a hundred years ago—also stood at this juncture of the medieval and the modern.

The fateful year, 1492, both the year of the meeting of two very different cultural groups and the year of the expulsion of the Jews and Moors from Spain was, for bad or good, a major watershed in our and the "New World's" histories.

Jeffrey Russell has shown in his *Inventing the Flat Earth* (1991) that it was the era of the quatercentennial that invented the courageous Columbus who challenged the flat earth theories being claimed for the times by nineteenth-century pro-evolution scientists, whereas the quincentennial Columbus has become the great bringer of disease and cultural oppression to the Americans, both radical inventions by which we read our present.

Why this hyperawareness of our transformability? How can we so easily invent, deconstruct, and then reinvent? My approach to this radically fluid postmodern context is to look at our lifeworld, which I contend today has a deep *technological texture.* For unlike the false but widely claimed and believed aspects of *science,* which takes itself still in a kind of modernist guise as universal, atemporal, acultural in its results, there has never been any doubt that *technologies* are more closely linked to both practicality and cultures.

I trace this subtle, often virtually invisible impact of technologies upon the lifeworld through all this first group of essays. And in the process I shift what has been the often standard interpretation which locates the "power" of technologies in older *economic* or *production* concepts (related to Weber and Marx) into a much more *cultural* perspective. Technologies are seen as "cultural instruments" which are non-neutral and deeply embedded in daily life praxes.

Among these technologies, just as optics was for early modern science the *image technologies* are for the postmodern context, and it is these technologies which provide clues for the multiple "readings" essential to postmodern perspectivalism. Indeed, the fascination of postmodernists with "virtual reality" experiments and developments indicates something of this recognition. Virtual reality, image technologies, and other communication technologies give distinctive shape to the postmodern context.

I conclude this first part with a rather high altitude overview in the chapter on "Technoscience and Pluriculture," which gives clues to the questions of how and why we today tend to read ourselves in such fluid ways.

1

Leonardo: Vision and Technology in the Renaissance

During the fall term of 1988 I lived on the flank of Monte Albano, overlooking the Florentine Valley and just around the corner from the tiny village of Vinci, writing a book on technology and the lifeworld. Little did I realize, when I first chose that location more for an appreciation of Italian cuisine and isolation from Stony Brook than for what turned out to be a more important and serendipitous reason, that this environment would be so important for this topic. This proximity to Leonardo's birthplace and the entire geography of what was the birthplace of modern science lent itself to meditations upon that Renaissance phenomenon.

The little museum dedicated to Leonardo in the village of Vinci became a repeated tour as I brought guests there—and I saw again and again models of his machines—Renaissance technologies, as it were. But as I examined the models and the drawings, it could not take much critical insight to quickly know that (a) most of Leonardo's machines, taken as anticipation of the future, were never built, and (b) if built, would not have worked, and even had those worked which could have, (c) would have been bulky and largely impractical.

For example, his design for flying machines—touted as anticipations of helicopters and airplanes—could never have left the ground.

The "helicopter" was a heavy, wooden screw, and his airplane, a wooden and cloth set of birdlike wings, sometimes not very different from the actual models made at the end of the nineteenth and early twentieth centuries, which you have all seen in film with amusing crashes and collapses. The design level of Leonardo's machines were basic applications of ancient physics, with combinations of screws, levers, inclined planes, winches, etc.

Why, then, the fascination with Leonardo's *imagined technologies?* That is what they were: exercises, sometimes brilliant, sometimes anticipatory, but usually only imagined. Indeed, our fascination with Leonardo's technological drawings is almost entirely *a twentieth-century phenomenon.* It began with a National Exposition of the History of Science in Florence in 1929 and the construction of some models of Vincian machines. Kenneth Clark did a study of Vincian machines in 1935. And in 1939, on the eve of the march of fascism, a major exhibit of such models was produced and later bombed when on a visit to Japan. (The irony concerning technology here should not be missed!)

There is something very modern about da Vinci and technology, to be sure. Once he decided that he had gotten all he could from Florence and fixed upon a move to Milan, Leonardo—in a darker anticipation of another connection between modern science and technology—wrote the Duke, Ludovico il Moro, a "grant proposal" for support in that court:

> I know how to build very light strong bridges, made to be easily transported, so as to follow and at times escape from the enemy, as well as others which are safe from damage by fire and from battle wear, easy and convenient to take apart and build. . . .
>
> I know techniques useful in invading a territory, like how to drain the water out of moats and how to make an infinite number of bridges and covered walk-ways and ladders and other machines useful for such expeditions. . . .
>
> I also know how to build cannons which are easy to move, that can shoot a small storm of small stones and create smoke which heartily frightens the enemy.[1]

But there is also something more ancient in da Vinci's fascination with machines. He was actually elaborating, and turning into a new and unique form of art, a tradition which was already nearly two centuries old. In the mid-thirteenth century, Roger Bacon had claimed:

> It's possible to build vessels for navigating without oarsmen so that very big river and maritime boats can travel guided by a solitary helmsman much more swiftly than they would if they were full of men. It's also possible to construct machines for flight built so that a man in the middle of one can manoeuver it using some kind of device that makes the specially built wings beat the air the way birds do when they fly. And similarly it's also possible to build a small winch capable of raising and lowering infinitely heavy weights. . . . it's also possible to build devices for walking on seas and rivers and for touching their bottoms without taking any risks.[2]

This was a tradition Leonardo knew well. Indeed, many of his machines are sketches of precisely the machines mentioned. They are the mechanical side of other fantasies of the times, such as those of Bosch's paintings. Leonardo, in true Renaissance fashion, was imbued with what could be called the *technological spirit.* He "imagines the world as a giant machine set in motion by spiritual forces and controlled in its perfect mechanisms by a superior intelligence that has arranged everything according to mathematical laws. . . . For him it's possible to reach an understanding of the 'infinite reasons' of the Universe, the mechanisms of the great machine of nature."[3]

Even the universe-as-machine was a tradition from the time of Roger Bacon. Lynn White, Jr., the great historian of medieval technology, pointed out that the use of the clock—which has become commonplace by the thirteenth century—had already become the metaphor for nature itself:

> It is in the works of the great ecclesiastic and mathematician Nicholas Oresmes, who died in 1382 as Bishop of Lisieux, that we first find the metaphor of the universe as a vast mechanical clock created and set running to God so that "all wheels move as harmoniously as possible." It was a notation with a future: eventually the metaphor became a metaphysics.[4]

So, if Leonardo, thought since the twenties to have been the great anticipator of a contemporary future, is also the continuer of an older past, and if I have now situated him as between the ancient and the modern, what does that tell us?

It tells us, first of all, that for a long time, a time longer than the rise of modern science in the Renaissance, European-Western civilization has been fascinated with the machine, with technology. Leonardo's world, his *life*world, was both permeated with burgeoning technologies and textured according to technical feats both celebrated and commonplace. I have argued that this "technological revolution" which occurred within the medieval period was the already sedimented background necessary for what was to become the modern scientific lifeworld which took its more distinctively scientific shape with Galileo, nearly a hundred years after da Vinci.

The engineering marvels of the Florentine Duomo—and Leonardo's suggestions about building were equally ludicrous to the builders of the time—took place within a culture where the clock had already regulated social life for centuries, where the invention of the cannon had closed off feudal life and the science of ballistics already had been invented, and where, only a couple of years after Leonardo appealed to il Moro, Columbus was to use the mathematics of global and celestial navigation to discover the New World. Modern science was born *in the midst of an already technologized* lifeworld.

My purpose is not to demean or belittle da Vinci, although it is to locate him in a milieu and to place him between the already ancient European love of technology and the modern fear and fascination with its contemporary results. But if Leonardo merely took his place within a worldview which accepted the mechanical metaphor of nature itself, understanding that the language of that world was mathematics—both to be elements of the century later Galilean science in Pisa—what did he contribute?

The answer lies closer to that other dimension for which we more primarily know Leonardo—his *visual* art. And his visual imagination. Leonardo's contribution was to *transform* and to *reformulate* perception itself. But, interestingly, the clues for this transformation may be better found first in Leonardo's anatomical, rather than technological, drawings.

The first move, retrospectively, seems almost too obvious—it is the move *to* vision, but *from* the other senses. The Renaissance was to be the enhancement of the visual/the reduction of the other senses. This shows in Leonardo's anatomical drawings. The anatomy texts of the time were replete with references to tactile qualities, to an anatomy conducted with-the-hand on the dissecting table (itself an activity not of the highest and most approved sort). Leonardo takes the corpse off the table, away from the hands-on dissectors. He distances the corpse from tactility and places it at a distance with what today is a visual representation of the various blood vessels, muscles, and other details of the body. He peels it back, layer by layer, in analytical fashion—but visually. This visualism is not simply an enhanced, a hypervalued visualism. It is also the simultaneous creation of a certain kind of distance. The Renaissance, with its long celebrated invention of the "laws of perspective," not only takes the actors out of the paintings and places them at a distance, but it makes of the subject of the art a kind of *spectacle.* It is a kind of visual invention of what will only much later be identified as "objectivity." In its own context it is a fascination with both a microcosm and a macrocosm which, visually, at least interpreted "geometrically," is the same phenomenon. Leonardo's is a particular kind of visualism which initiates the modern.

One can, furthermore, see the parallelism instantly between the body and the machines he draws—the same analytical detail, the same focusing in on joints/hinges, organs/gears. And part of this technique is to become a mainstay of engineering itself—*the exploded diagram.* We see it in the representation of a uterus with fetus; and equally, in the diagram of a gear-propelled vehicle.

This is visualist thinking, the very thinking which would become the style of thinking in both modern science and, even much more particularly, in modern *engineering.* It is the exploded diagram which makes the workings of such inventions as diesel engines, as hydraulic pumps, as the instrumentation of science itself understandable. These are the visual languages of engineering.

In a more general sense, da Vinci also fits into the already established convention of three-dimensional, perspectival drawing which is associated with the invention of the laws of perspective, sedimented by 1440, prior to Leonardo's birth. These, Leonardo brought to a new level of perfection. The "bird's-eye" views of cities from a nonoccupied, but only imagined, bodily position, the convergence at infinity, and the preoccupation of the Renaissance with the microcosm and macrocosm—to be sure, in theater and spectacle—were the beginnings of the invention of the *observer.*

This is also why, in the already established metaphor metaphysics of the great machine of nature, the anatomical and technological drawings are, at base, *alike.* It is the prelude, the necessary condition, for the coming age of technology.

Much, much later this would be recognized by the twentieth-century thinker Martin Heidegger. It was he who noted that technology is not some collection of artifacts but a "way of seeing." And it was Leonardo da Vinci who helped to make that "technological" way of seeing into a gestalt which could be refined and extended, even to the current and still perspectival representations that today capture our fascination in the now artificially produced virtual "worlds" of our latest toys, computer graphics.

There is thus not some mere juxtaposition of the Renaissance and technology; there is, rather, the very most intimate synthesis of a way of seeing which was the birthplace of the modern.

2

Columbus: New Technologies/ Old Cultures

A *quincentennial* is a rereading of an event that is necessarily situated in both nostalgia and ambiguity. Today's world is one marked by the emergence of a global, highly technologically textured lifeworld perched—not always comfortably—atop old cultures.

The Columbian voyages from 1492 to 1504 were events that retrospectively are seen as crucial turning points in "our," that is, Euro-American, culture. The opening of the New World appears as a virtual historical anomaly, but one that sets the stage for what was to become the dominance of Euro-American technologized civilization.

The anomalous quality of the event may be seen in both Columbus's own failure to recognize that he had discovered a New World—he died believing that he had voyaged to just off the Orient—and the equal absurdity of his luck, since he predicted a landfall some three thousand miles to the west of Spain. He had believed Paolo Toscanelli's estimate of the earth's size, which placed Japan approximately where the New World turned out to be, against the wiser scholars, who correctly thought the earth to be about eleven thousand miles around.[1] But neither they nor he suspected the existence of the two vast continents of the New World. So this daring, false-belief-situated voyage may be seen as the expansion of the European world at the same time that the Renaissance began to open the intellectual world of Europe to what was to become a science embodied in material technology.

Why should Columbus's voyages be nominated for that watershed? We know today that he was not the first European to discover the new lands. For even if we dismiss the legends of Brendan the Navigator, who in the fifth century may have voyaged as far as Iceland, we do know that Leif Eriksson *did* sail to Newfoundland and actually settled for a time amidst the Skraelings at the turn of the millennium.[2]

Yet neither Brendan nor Eriksson can lay claim to the establishment of the exchange that became permanent through the voyages of Columbus. His voyages carry with them the continuity of both memory and communicative exchange which are culture shaping. In our history, they were monumental, or so we think as we look back retrospectively. We see Columbus as the harbinger of, first, the colonization that took place on the part of Europeans in the New World and, then, the dynamic, four-century development of both science and technology, which transformed both European and American culture and which today has taken the entire globe as its resource. Brendan and Eriksson stand only as episodes and belong to a fringe of our historical memories.

This cultural current was one that swept over all other cultures. And even if the two and a half months that it took Columbus to reach the Caribbean is now spanned routinely in a third of a day, the event began what is today's Euro-American cultural dominance.

We tend to think that science and technology are distinctly Western in origin and development, and, insofar as they dominate the globe today, we could be lulled into thinking of a kind of colonial victory over the earth. But the outward current carries under its surface a countercurrent, and if the surface is that of a dominant new technology, its undercurrent is that of many other old cultures. What does this countercurrent portend? I shall examine this question by returning to voyages, not just those of Columbus and his European competitors, but of two very different cultural histories of a more ancient form.

The first group of such voyages lies in a legendary, forgotten "history," that of the Pacific Islanders. Today, we know that virtually every inhabitable island of the Pacific—with its far vaster reaches of sea than the smaller Atlantic—was reached and settled by Polynesians, Micronesians, and other islanders long before Leif Eriksson reached Vinland. We know, too, that once settled, these islanders remained in contact and exchange with one another through periodic voyages of pilgrimage and trade undertaken in the same multihulled boats that characterized the distinctive naval technology of the Pacific. What is amazing about these feats is that they were undertaken with technologies that were "stone age," insofar as they did not employ metals, and by preliterate peoples, insofar as the cultures were oral ones.

The second group of voyages—as so many of our antecedents turn out to be historically—was Chinese, and they preceded Columbus by slightly less than a century. These were the voyages of Cheng Ho from 1405 to 1433 on behalf of the megalomaniac emperor Yung Lo. These voyages opened up much of the Pacific and Indian Oceans and reached to the very back door of Europe itself in the Persian Gulf.[3]

As navigational feats and geographic explorations, both of these sets of voyages easily equal that of Columbus. Does our ignorance or disregard of such voyages simply bespeak our own cultural chauvinism? Or does the Columbian venture still remain of greater historical significance?

If we simply assume the superiority of our Western technology, we might be able to conclude at least that neither of the two earlier, spectacular voyages succeeded in spreading either their technologies or their cultures beyond the limits of the originating source. Yet there are ironies here, too. And I shall look at these by first focusing briefly upon the naval technologies of the three sets of sailors and their techniques of long-distance orientation, or navigation.

I begin with the most ancient of the three sets of voyages, those of the Pacific navigators. By Western standards, their craft surely would be judged to be the most primitive of the three. Their "low-tech" multihulls were wooden variants upon hollowed log canoes, stitched together with coconut fiber ropes, caulked and sealed with vegetable tars, and powered by crudely woven materials. Similarly the instrument-fascinated Westerner would be convinced that the totally instrumentless navigational techniques used by the islanders also should be thought primitive.

Such a judgment would be both superficial and chauvinistic. For, in design and function, the variety of ancient multihulls was seen, even by the early European explorers, as having certain superior features to their own slow and hard-to-maneuver vessels. The larger of the seagoing catamarans of the Polynesians were recognized by Captain Cook to be from two to three times as fast as his ships. As they were unballasted, these sailing vessels were also unsinkable. The multihulls, with a relatively simple crab-claw sail, also were more maneuverable than the square-rigged and more complex ships of the Europeans. Indeed, today "high-tech" versions of the multihulls have broken major speed records set by earlier sail craft, including those of the clipper ships, whose waterline length exceeded that of the multihull racers by factors of three or four. What should be historically amazing in this is the failure of the Europeans to adapt this design technology and to turn it into speedy vessels of commerce or war, as they did with so many other technological borrowings from other cultures.

Similarly, the instrumentless navigational techniques of the islanders—only seriously studied by Euro-American anthropologists in the middle to late twentieth century—not only were obviously successful, even in locating new islands out of sight, but were a cultural marvel. These navigators learned to use the stars, wave patterns, bird paths, atmospheric reflections: in short, a most precise and subtle reading of natural phenomena, in an oral perceptual system not exceeded in accuracy by any non-Pacific contemporary navigation schemes.[4]

I shall not go into great detail here, but shall note that, without a compass, the islanders used star motions and parallax phenomena to orient themselves as they moved across the ocean. Swell patterns were used for course direction, and even in storms expert navigators could detect the swell rhythm under the surface confusion. Bird routes bespoke nearby but unseen islands, as did the reflection of foliage upon permanent cloud masses about island mountains. Not believing that a perceptual navigation system could be so used, European experts as late as 1957 surmised that old, worked shells with two holes must have been used by Polynesian sailors as primitive wind speed gauges when, in actuality, the two holes were for the thong that made the shell into a pendant for a necklace. The European assumption is that instruments must mediate controlled interactions with nature.

In studying the Chinese navy, sent forth by Emperor Lo, one sees that the disparity with Columbus is even greater. We recall that Columbus had three small ships, the largest of which was the Santa Maria at about 98 feet. Admiral Ho's fleet numbered 317 ships, the smallest of which was the combat ship at 180 feet, and the largest, a *nine-masted* treasure ship of 444 feet! His crews numbered 37,000 men. The ships also were of highly sophisticated design, with waterproofed separate compartments, balanced rudders, easily reefable, junk-type, fully battened sails, and were large and even multistoried.[5]

We know that in navigation Admiral Ho used the compass, which was after all a Chinese invention, and that he had elaborate charts of the then known world, with compass bearings and the use of other directional instruments. His seven voyages took him as far as Java, Sumatra, all around Indian Ceylon, to the Maldives, and up to the Persian Gulf.[6]

My point so far is simple: naval technology and navigational techniques of the pre-Columbian voyages were at least as good as, and perhaps better than, those of Columbus. The distance and exploratory accomplishments of these voyages were as startling as those of Columbus, or more so. Yet, it still may be claimed that neither of these epoch sets of voyages mark the juncture between an Old World and a New World epoch.

In part, the reason for this lies in the cultural settings that determined the ultimate result of the two non-Columbian voyages. The earliest voyages of the Pacific Islanders, like the ancient pasts of all oral cultures, lie beneath clouded legends. We do not know when they first occurred, only that they succeeded in colonizing previously unpeopled islands. We do not know the actual motivations for doing the exploration. Some surmise that population pressure or tribal wars in which losers had to leave motivated the leap into multithousand-mile voyages. But because the islands were uninhabited and because the voyages did not reach the inhabited mainlands of the Americas, the cross-cultural and cross-material cultural exchanges that followed Columbus did not occur. The voyages were, so far as modernity is concerned, forgotten.

In the Chinese case, the history and motivation is much clearer and even finds some very contemporary echoes. The Emperor Yung Lo, not the first in his history, believed Chinese culture to be superior to all other cultures. His fleet was to express this superiority by traveling to the far corners of the world and, through gift-giving and receiving, to proclaim this superiority. That was one reason why the treasure ship was the largest in his fleet and the combat ship the smallest.

On reaching a port, his ambassadors would march to the local potentate and give elaborate gifts, proclaim China the center of the world, and seek trade and homage treaties, which in many cases they got. Yet, the return in homage and trade was not enough to balance the cost to the treasury of the gifts and of the proclamation of superiority.

By 1433, Admiral Cheng Ho's voyages were ordered stopped, and in another recognizable reversal so typical of China, echoed again in recent events, foreign travel was now forbidden by Emperor Lo, and China turned inward upon itself.[7] To voyage, or even to build a long-distance ship, became a capital offense in the decades following the brief outward movement of Cheng Ho. Thus, while not historically forgotten, the consequences of these Chinese voyages were reversed.

Does this leave Columbus, almost by default, the voyager to have turned the world from old to new? If so, it was clearly not because his outward voyage was one undertaken with either a superior naval technology or superior navigational technique. Of course, one might argue—and this argument has often enough been made—that what differs was the rapacious and hungry culture that Europe had become and within which Columbus's voyages must be located.

Lynn White, Jr., the preeminent historian of medieval technology, has argued that it was the West's insatiable hunger for power that transformed its borrowed technologies into instruments that produced work and control over nature.[8] Thus, the Hindu prayer wheel, adapted, became the windmill that drained the marshes and claimed lowlands of Holland. The stirrup adapted from ancient Mongol designs became the platform for the early form of a battle "tank," the mounted knight with horse powered lance for forward thrust.

Daniel Boorstin, who recalls the Chinese voyages, notes with irony that the Western explorers surely would not have been satisfied with ritual recognition of their superiority. The conquistadors who followed in the Columbian wake simply usurped the Native American lands and realms and paved the return to Europe with a trail of confiscated gold.[9]

My point here is not simply to decry the greed and crudeness of our ancestors, although one cannot deny it. My point is, rather, to show that technologies in the ensemble are distinctly embedded into cultures. This is one of the reasons why the previous focus upon the naval and navigational technologies and techniques did not reveal the distinctiveness needed to account for the watershed quality of Columbus's voyage over that of the islanders or Cheng Ho.

What is needed is a deeper insight into the ways in which the ensemble of technologies relates to cultural gestalts, particularly those of perspective adopted and of an implicit view of nature. In the case of Columbus, the perspective was set early by the literate culture of the West.[10] Interpretation was largely through seeing the world as text and, in the case of navigation, through reading charts. Portolans, or charts detailing a coastline with dominant wind directions, had been common for two centuries or more prior to Columbus. He also knew the mathematized way of seeing the world embedded in the concepts of latitude and longitude and knew that, if he followed the same latitude westward, he would reach the Indies. But more important, with his mathematical seeing, the crude astrolabes for determining latitude, and, particularly, the ability to assume a fixed overhead position looking down upon the map that represented flat earth, he could plot his place upon the spherical earth. I claim this overhead or god's-eye view is deeply embedded in Western culture and is indicative of some of the other features of dominance I have already noted, which belong to the way of seeing typical of Columbus's as of later times.

In deep contrast, the islander navigator did not even conceive of the earth as a globe, but as a "mother" who would bear him to the distant lands in her care. Lacking both the practice of text reading and its overhead view, the Pacific navigator instead followed a practice of learning oral songs that carried the formulas for voyages and a body relativism for calculating apparent directions. The perspective adopted was one of embodied "being-here." Once on board, the navigator takes his actual bodily position as constant and judges all relativistically according to that reference. Even the language changes, and one says, "Tahiti is approaching . . . " rather than "I am nearing Tahiti." There is no bird's-eye view from above that would allow one to measure a distance between sailor and landfall. This embodied way of seeing is consonant with the way of seeing oceanic nature characteristic of that culture.

If there are, indeed, two very different ways in which technologies and techniques are culturally embedded, the question might well arise as to which is superior. There is no simple answer to that question. If we take today's standards and anachronistically apply them, we might be tempted to say that, in terms of accuracy and precision, the Western mode is clearly superior. Today's geosatellites, LORAN, and computer based navigational systems can locate users to within a few feet of their actual position. Such accuracy, however, contrasts just as sharply with the dead-reckoning abilities of Columbus as with the islanders, and it is clearly an unfair standard of comparison.

Nor is it possible to compare fairly the mathematically enhanced and interpreted systems of today's developed technologies with the lack of development of a perceptualistic and relativistic set of possible technologies that could perhaps equally have enhanced the islander mode of navigation. It is only very recently that we have begun to develop what are called *virtual reality* systems, which enhance bodily perceptual capacities to the same degree that our digitalized and mathematized systems do.

Trainers for pilots that simulate actual flying conditions, and fog-erasing displays for actual landings that make runways visible, are preliminary examples of a virtual perceptual system that would enhance a bodily perceptual orientation in space just as well as the already developed system for mathematized, map-reading systems. The fact that one system is better developed today than the other is more an indicator of a sedimented preference of a single tradition than of any absolute technological superiority of one approach to the other.

These very observations, however, point to something much deeper than appears on the surface. They point to a phenomenon of the late twentieth century about a different change relating to new technologies and old cultures.

At the beginning of this itinerary, I hinted that, while it seems obvious to all that the dominant current sweeping the globe in the twentieth century is the spread of Euro-American technological culture, there is no culture, primitive or complex, local or continental, that has not been touched by technologized civilization. Indeed, one of the primary tasks demanded of all cultures is to respond to that current, be it by resistance, modification, hurried acceptance, or supersession.

This process is uneven at present. There are areas where only small parts of modern technology have been adopted, while others have become so modernized that even Europe and the Americas seem doomed to being superseded. (The rush to modernization in parts of the Pacific Rim—such as Korea, Japan, Singapore—is so dramatic that when I return from such places to New York, it seems that I am returning to a decaying civilization.)

I suggested, however, that simultaneous with, and inextricably related to this dominant outward moving current is a countercurrent that returns underneath but along with the top current. That is the countercurrent of the Others. For every contact the Euro-American technologized culture makes with the Other, there returns a countercurrent of the culture contacted. This is the phenomenon of what I shall call *postmodern pluriculture.*[11] It is superficially evidenced in such phenomena as fashion, cuisines, and popular culture. Fashion today is a bricolage phenomenon. It arises not only in the old centers of New York and Paris, but also in Tokyo and Rio. Themes may be Japanese, Spanish, paramilitary, or tribal. But the culture bits and pieces that make up whatever dominant and recessive fashion statements are current are the flotsam and jetsam from the Others of whom we are now aware.

The same may be said of cuisines. Every major capital now has, almost mandatorily, a plethora of cuisines. There will be variants upon Oriental cuisines, both generic and specific continental cuisines, and some variety of others, often reflecting the trade and taste patterns of a capital. Like fashion, the plurality of cuisines is the backflow from what was the dominant exploratory outflow of Western technological culture.

Popular culture is yet another variant upon the multicultural phenomenon. While jeans may be universal and rock music international and while the elements of that culture are often seen as dominantly American, its roots and development are never simply that. The MTV video is among the most radical of the bricolage phenomena, with its fragments and culture bits the very essence of the short-lived audiovisual presence. Last fall, while I was writing in a very tiny village in northern Italy, the local bar from which I had to make my weekly international call was the center for the local teens to absorb this popular cultural initiation. Here, too, was pluricultural bricolage.

It may be objected that this fragmentary incorporation of multicultural bits is too superficial to characterize the sense of an era. But I have already hinted at a deeper manifestation of the same phenomenon. It lies in the comparative study of the three sets of navigators we have just studied. It belongs to the late twentieth century to have analyzed and in some degree appropriated the different sets of principles and techniques that originally belonged to only one culture. Ours is a pluricultural vision. Thus, if today one wishes to maximize speed in sailing, one builds a high-tech version of an ancient Pacific catamaran or trimaran, rigs it with a wing or high-aspect Bermudian sail, and proceeds to exceed the actual wind speed on certain angles of sail. Such a racing vessel owes its concept to a multicultural technological past. The same obtains in the multiple—we say, redundant—navigational systems that we incorporate in flying or submarining or sailing. Today, we alternate from the fixed overhead positioning of the older European map-reading techniques to the relativistic and embodied positioning of the Pacific Islanders, reading nature and LEDs.

This bricolage, multiple seeing, is a seeing that is more distinctively a development of the late twentieth century. It is the inheritor of our practices—if not yet of our believable ideology—of cultural relativism. It is the seeing with what I call the compound eye. That vision is never single. It is a series of multi- and alternative visions, symbolized in the growing presence of the multivisual screens we have become familiar with, such as those found in all television newsrooms, in control centers for large plants, and even in department stores. Compound vision is multiple vision. One scans the multiple screens, focusing here, then there and, out of the mélange, forming new directions and possibilities.

What we perhaps have not noticed is that this new vision, relativistic and compound, is both Western and yet post-Western. It is Western in that it arises and takes place in the matrix of a Western-originated technological culture, but it is not Western in that what have been the foundational and dominant ideologies we have taken for granted now find themselves placed alongside other, distinctly non-Western beliefs and ideologies. This is frightening to some, but exhilarating to others.

Of course, it is not the case that such multicultural phenomena are entirely new. Spain's history, for example, is a witness to pluriculturality, at least in the sense that it became what it was by the combining and blending of Moorish, Jewish, and Catholic cultures. History perhaps always has been driven by multiculturality, but our contemporary compound vision enhances our appreciation of this plurality and raises it to a prominence not previously noted. But it also, in its relativistic tolerance, refuses to impose or colonize that which is Other in its midst. At least, that is the secret hope lying within a compound vision.

So, just as the Jewish mathematicians located in the school of Henry the Navigator, the revival of ancient Greek science through the Arab philosophers, and the universalism of the Christian religion became the basis of what was to become the world march issued from the Columbian voyages, so today we stand at the threshold of another new world, this time one that is the globe itself.

The shift in vision, however, is not one that belongs solely to a Western evolution. It is one that, while linking the globe into a single communications network, incorporates a diversity it has yet fully to recognize, let alone deal with in a synthesized fashion. Cultural terrorism takes its place alongside enlightenment-tolerance cultures, socialism finds itself accommodating to consumerism, and ancient religious strife takes distinctly nationalistic shape. The vision of the compound way of seeing remains, at the dawn of the twenty-first century, at best a bricolage of the pluricultural. But its world is a new world, newer and as different from the modern one it is now leaving as Columbus's Indies were from the Orient he had sought.

None of this is to say that pluriculturality in our technological civilization is without its own distinctive problems. But pluriculturality is what gives these problems their distinctive set. Indeed, many of the most problematic of our social-political problems may be seen as a response—whether utopian, negative, ambivalent, or resistant—to technological change, and as an enhanced sense of ethnicity and even ethnic conflict that inevitably accompanies this awareness.

For example, with the breakdown of previous socialist versus capitalist ideologies apparent between Eastern Europe and the West, there has emerged—or reemerged—a series of struggles concerning ethnic groups and decentralized regionalisms. Similarly, much of the intense strife within the Middle East simultaneously displays revived senses of religious fervor (Muslim fundamentalism, though all fundamentalisms display similar characteristics) in forms that react both to technologization and to apparent threats to traditional religious and ethnic values.

Postmodern pluriculture is essentially not neutral. The plural view is acidic in its effect upon all deep monocultural traditions, and clearly this is deeply felt within monolithic ethnic cultures and religions. Here the conflict of new technologies and old cultures appears as one of the distinctive problems of postmodernity.

It is too early to tell what the outcome may be. One may hope that a dystopian result, conflagration around new issues postdating the previous superpower conflicts, can be avoided. Yet, any simply utopian hope, like the ultimate resolution of religious intolerance resulting from the Western enlightenment, seems more difficult in today's large mélange. But like the secret acidity of the enlightenment, which solved the problem of religious tolerance only by making religions redraw the lines of truth from absoute and public at least to absolute and private, so may pluriculture have to hope for the same aim. This is to say that pluriculture also changes culture, and that will be, in one way or the other, the result of the meeting of new technologies and old cultures.

3

Technology as Cultural Instrument

The topic here is *technology transfer.* By that I mean the introduction of some set of material artifacts out of their original context of human praxes or techniques, into some other cultural context. Such transfers overall are commonly recognized as the spread of so-called "advanced" technologies, such as those common to the northern hemisphere high technology, high industrial nations, to the so-called "less developed" and largely southern hemisphere nations. On a smaller scale, technology transfer may also mean the adaptation of some single technology of one people by another through means of cultural encounters.

Technology will mean here, as it has in my previous work, some artifact or set of artifacts—material culture—related to a context of human action or praxes (which include techniques of use). I shall be focusing, however, upon the exchange itself and in particular upon its cultural dimensions.

When a "technology" transfer occurs it has long been known that more than a transfer of some artifact is involved. The steel axe has been the classic example of a simple transfer. The standard analysis usually goes something like this: (a) When a "technologically advanced" people—with steel axes—come into contact with, say, peoples using only Stone Age tools, there will be an initial flow of artifacts from the advanced to the indigenous people. They will simply adapt and even covet steel axes. (b) The reason, the standard analysis usually holds, is that the steel axe is obviously more efficient and functional—it cuts down trees faster and with less effort. Never mind that this reason is also ipso facto already an assumed value within the "advanced" culture. (c) But, then, at first unknown to the recipient people, more is being received than a steel axe. What, underneath, is being taken in is a new set of relations, primarily economic, which will eventually result in dependency relations for the indigenous people. Thus, because the indigenous people, now accustomed to the steel axe, cannot produce one—they do not have the technological praxis—they must enter trade and other relations (furs for axes) which eventually make them dependent upon the "advanced" culture.

This standard analysis is not wrong per se, but neither is it very penetrating and it is too simply skewed to economic factors. Heidegger, for example, has used the radio as an illustration of a technology transfer which comes closer to what I am after with respect to technology as cultural instrument. It is well known that two artifacts which are immediately fascinating to previously isolated indigenous peoples are wristwatches and radios. Nor is it necessary that the context of understanding which belongs to the watch or the radio be immediately apprehended. Fascination with movement, light, and particularly, sound, is sufficient to make the artifact desirable and to start a process of "culture transfer." And, indeed, one would probably be hard pressed to isolate two more crucial technologies for this purpose than the clock—which transformed the entire sense of time and public order in Western technological civilization—and the radio—which has transformed the entire communication and embodiment of language presence in that same history. But what the clock and radio transfer, as cultural instruments, is much more than a set of economic relations. Such artifacts are the non-neutral means by which an entire cultural gestalt can be carried.

Although I shall keep the narrative of this analysis concrete by use of cultural variants, there are certain theses which I am arguing which may be placed in the open.

Thesis One. There are no neutral technologies, or, positively put, all technologies are *non-neutral.* I use this phrase deliberately to eliminate any tendency to prejudge technologies as simply "good" or "bad." They are transformational in that they change the quality, field and possibility range of human experience, thus they are non-neutral.

Now in one sense, this is a trivial insight known at least intuitively by most people except for certain philosophers, philosophers of science, social scientists and the older fashioned scientists. Industrialists, engineers, commercial people, pioneers and militarists have long known that technologies are non-neutral, as have some philosophers in certain traditions.

This thesis, however, becomes clearer if one is also aware that technologies must be understood *phenomenologically*, i.e., as belonging in different ways to our experience and use of technologies, as a human-technology relation, rather than abstractly conceiving of them as mere objects. Were technologies merely objects totally divorced from human praxis, they would be so much "junk" lying about. Once taken into praxis one can speak not of technologies "in themselves," but as the active relational pair, human-technology. The human with a steel axe is different than the human without one—the transformational effect becomes clear when we regard as the *primitive* of our analysis, this *human-technology pairing.*

Thesis Two. If abstracting technologies as neutral objects is one error, then reifying Technology (with the capital "T") is an equivalent one. There is no unitary, determined single destiny to technological development. There is a multistable and diverse and ambiguous set of multiple directions whose ends are probably not predicable any more than any historical-cultural development can be adequately predicted. Thus I am arguing against the Elluls, the Marcuses and other thinkers who claim that there is a single, inevitable direction to a Technology which is out of control. The "power" of technologies, conversely, must be isolated at a more subtle level, that of human social and cultural interchange, examples of which I shall examine.

Thesis Three. The dimensions of technology transfers are never simply economic or productive, but multidimensioned and involve basic cultural and existential interchange. This is therefore a rejection of any foundationalist or reductionist explanation and an opting for a more multidimensional and phenomenological model of understanding. It will involve utilization of a *variation theory* such as originated with Husserl, but now adumbrated into the historical-cultural domain both through interdisciplinary use of historical and anthropological insights and of imaginative variants upon these. In this sense I may be thought of as developing a non-Marxian thesis regarding interpretation, without being anti-Marxist.

Example 1: Gold Prospectors and New Guinean Highlanders. In 1930 the first white colonists from Australia began to penetrate the New Guinean highlands in search of gold—they were totally unaware that they were to encounter multiple tribes of highlanders, now estimated to have been nearly a million in population at that time. Although not with scholarly purposes, the prospectors fortunately took a movie camera with them and recorded these first encounters—a recent, remarkable educational television program played this film and also provided interviews with older highlanders who remembered this first encounter. I shall focus upon certain technology transfers which occurred in these first meetings.

It should be pointed out that at first, most of the highlanders regarded the newcomers as either *gods* or *ghosts*. In the recounts, they emphasized that they had never seen *white* people and that their cultural beliefs held strongly that they (the highlanders) were the *only people* living in the *only world*, the highlands. A plausible notion, then, was that the white newcomers were ghosts or spirits from some "beyond." I need not point out that this pattern of shock and interpretations is, far from being unusual, almost typical of first encounters between peoples. It, in fact, echoes some of the first encounters recorded between Spaniards and Incas in Latin America.

Later, of course, they gradually discovered that the newcomers were mortals—interestingly in quite concrete and perceptual ways. Because the colonists wore clothing, the highlanders at first thought that ghosts or gods did not defecate, fornicate or urinate—but soon someone spied a newcomer doing his duty and as the film showed, every material leaving was immediately investigated. The highlanders soon discovered that the excrement had the proper *human* smell! One could say, a very "earthy" discovery!

Now, however, focus upon the technologies: There were, of course, steel axes and knives and these were immediately seized upon as fascinating and desirable. But, I suggest, not simply because they were functional or more efficient—axes and knives were *familiar* and *understandable* artifacts, already there in the indigenous culture. The point of juncture was *familiarity within a known praxical gestalt*. Steel axes and knives were different in material and appearance, but clearly easily recognizable and understandable. Contrarily, even though rifles were clearly much superior in weapon power than axes or knives, there was not much of an initial fascination with guns (until later). To demonstrate their power (again quite aware of the non-neutrality of the human-technology symbiosis) the prospectors killed a pig at close range. There was a shock of fear at the report of the gun, but as the film showed and the reports indicated, such a killing did not at first seem so impressive. Anyone could kill a pig at close range. Later, after the tribesmen came to realize that the whites were mortals like themselves, their *technologies* were perceived as not only fascinating and desirable, but perhaps capturable—so a raid was prepared and occurred with the predictable, disastrous result. Nine islanders were slain by a fusillade of riflefire and only *then,* because this occurred at a distance beyond spear range did the new technology become awesome.

Were I to leave the example here with the axe-gun technologies, the standard analysis might seem not only correct, but sufficient. But that was not exhaustive. The colonists carried tinned goods, for example those old elliptical sardine cans. These they discarded along the way as "junk." But the lids were immediately picked up and *incorporated into elaborate headdresses* by the highlanders. Again, there is a pattern of a new artifact first being incorporated into a known praxis, a fashion praxis if you will. And, again, for those who have known all along that fascination and adornment are as good a way into the hearts and minds of humans as any, recall the long history of trading which has involved cheap and gaudy goods in initial encounters. The Spaniards carried vast quantities of hawkbells for trading (shiny and sounding). And, in the cases mentioned, one interesting fact is that the bauble, if it had some original use as a hawksbell, is taken into an entirely new and different use, but one which is familiar to the recipient. It is not far from hawksbells and shiny tin cans to television and, one might add for some fascinated with today's screens, videogames and word processors!

My point is this: empirically, the steel axe and the mirror are simultaneously and equally fascinating and desirable. It takes a peculiar and selective perception to ignore the latter and choose the former as basic, which is to say that a simplistic economic determinism is already a chosen theory which is then self-verified. Instead, I am suggesting, there is a structure, a multidimensioned structure, to such interchanges which simultaneously involve a multiplicity of levels and functions.

The interstices which initially occur, take place in certain crucial familiar praxes: axes and knives to familiar functions, mirrors and baubles to other ones. Thus we find the strange anomalies of certain artifacts used in ways entirely different from their designed purposes. Watches, for example, while desired are not at first used to "tell time" since the whole content of hours, minutes, and seconds is initially unfamiliar. Instead, they are used for decoration, prestige, and for a kind of mini-entertainment through their noises and motions. A more functional example comes from the use of a compass by the Puluwatians in Micronesia—although used for steering a straight course (and thereby replacing the more subtle reading of wave patterns of traditional navigation)—it is not used to tell points of direction or degrees (steer 180-degree course) since this system of navigation remains foreign to the traditional navigators of Puluwat.

There is a deeper significance here than may appear. It is two-sided.

1. First, any technology may be used in a multiplicity of ways, limited only by the individual and cultural imaginations of the people or user. Heidegger's famous hammer is a good example. In use it is not an object-as-such, but the transparent and familiar attainment of a practice. But, equally, it may be used in any number of ways. It could, and perhaps is dominantly used, for its designed purpose—to hammer. But it could be used as a paperweight, an objet d'art, a murder weapon, a pendulum weight, a door handle, etc. This ambiguity of uses is, however, not indefinitely extendable. The hammer would not serve well as a rope, a sewing needle, or other artifact which in its range would also display some span of multiple uses. Here is a dimension to human-technology uses which is deeply unpredictable and uncontrollable. But it is not something which belongs essentially to modern technology; it has always been a dimension of human-technology relations. The cultural variations alluded to are simply dramatic examples of precisely this possibility structure. Obviously, such possibilities may be trivial, benign, or frightening. Even a complex nuclear plant *could* be used in a number of ways—for example as a giant grain silo, converted to coal, taken as a piece of junk or, better, junk sculpture, preserved as a monument. And its products, as we so frighteningly know, can be used for making terrorist bombs (nuke packs now come in the fifty-pound variety), poisons for the New York Waterworks, or any number of possibilities. No one can control these uses as such!

2. But if any artifact may be used in multiple ways, one might be tempted to say that it nevertheless "is," in some ontological sense, this or that thing. This is, however, to misconstrue technological artifacts. Apart from human-technology relations, such *technofacts* as they might be called, would at best be certain conglomerations of material being, for example their physical or chemical properties. But the being of a technofact is more like that of an art object than a natural object, not only by being made, but by its use context. Thus the ambiguity of uncontrollable, multiple uses for any single technofact is balanced by the ambiguity that I may also use any number of technologies to achieve the same purposes.

If I wish to hammer, I may normally use a hammer in its Heideggerian sense—but I could use a crowbar, a piece of steel sculpture, a properly shaped stone. This is not to say that the ambiguity of being able to use a multiplicity of technologies for the same purpose means that technologies are either neutral or irrelevant. To the contrary, it is to say that each variant upon a human-technology symbiosis has related possibilities—while each of the above technofacts could be used to hammer, the use of the palm of my hand or a waterballoon for this purpose would be futile. Yet the range of ambiguity in both multiple uses of a single technology or single purposes utilizing a range of technologies once again exemplifies both the essential unpredictability and ambiguity of the human-technology relation, and the impossibility of "control."

It is now time to return to the theme of technology as cultural instrument. In the examples given so far, I have shown how technologies are first taken into recipient cultures through what may be called a recognition of the familiar. The steel axe and knife and the shiny bauble both have immediate, recognizable roles to play in the indigenous culture, even if the recipient has no idea what he or she is "buying." The fascination and desire will be enough to ensure the giver or bringer of the technology of a wedge into the recipient culture.

However, there is another side to familiarity in the encounter between two alien cultures, and that is strangeness. Strangeness and familiarity are two sides of the single, multidimensioned phenomenon. Strangeness seems at first dramatic and obvious—but it usually lacks comprehension. In the example I have given, the taking of the gold prospectors as ghosts or spirits, there was a dramatic recognition of strangeness, and an attempt to bring it into familiarity—the concept of ghost or spirit. Let us now, however, turn the tables so that the encounter between two people will be seen to be two-sided. The colonists also perceives strangeness and attempts to familiarize it.

When the Columbians first began to explore the New World they recognized the strangeness of the peoples—there are remarks about costume (nakedness or near nakedness), customs (cannibalism and marriage practices among the Carib), and the like. But not unlike the highlanders, this strangeness is accommodated within a persistent familiarity. Not only did Columbus continue to believe that he had discovered the Indies or Orient, but the very category "pagan" is an attempt to place the indigenous peoples under a familiar idea, a religious conception not *functionally* unlike ghost or spirit.

What is especially striking is that strangeness is so difficult to perceive that representations of the indigenous peoples frequently remain distorted for long periods of time. For example, the first portrayals of "Indians" by early Colonial painters frequently look like Michelangelo anatomy drawings. To be sure, the feathers and loincloths are portrayed, but the physical features remain primarily European in appearance even if the bodies represented ones covered with body paint. In North America, for example, it was only in the nineteenth century that one began to get genuinely realistic representations, as in the extensive painting projects of George Caton. Interestingly, by the time portraits of this quality began to appear, it was already painfully apparent that the Indian way of life was disappearing. There remained, after some centuries of interaction, deep misperceptions between people of the different cultures. Yet what of material culture and technological interchange?

We tend to overlook that the exchange is always two-way, without denying that something in the exchange is not equal. The Spanish horse and later the rifle, completely changed the life-styles of plains Indians—in fact the romantic image of the roving hunters and migratory peoples which inhabit our imaginations were already postexchange social patterns. Similarly at the level of material culture, who are the Irish without the potato, the Italians without the tomato, the New Englander without corn, or the world without tobacco? All these were exchanges which came from New World material cultures. What is different between today and yesterday, is that all these products are now virtually universal in distribution. This fact alone both *diversifies* and *homogenizes* "World Culture."

That technology transfers will occur at cultural interstices, and that this exchange will inevitably carry culture change, has been illustrated. Furthermore, that the transfer will occur at some point of familiarity or fascination may also have been demonstrated. Moreover, these exchanges do not necessarily occur between variants upon low or appropriate technologies, but may occur between very high and very low technological contexts. India, for example, is attempting to "leapfrog" its rural development program by introducing television rather than a series of other technologies into rural areas. Thus the satellite receiver set up in a field being plowed by oxen becomes a symbol of India's development. The fascination of a village television substitutes for the hawksbell.

All this leads us, uncritically, to too frequently assume that there is some kind of overwhelming, even Frankensteinian trajectory to modern Technology conceived of with the capital "T." Yet without denying the march of the most recent technological revolution over the globe, two pauses should be taken. Are there modern technologies which are *not* adapted? And, if so, why? By shifting the emphasis, another aspect of transfer may be noted. Again, we can begin with early explorational examples.

The first European explorers entering the Pacific were amazed at the ship technology of the Islanders. Large catamarans, capable of carrying two hundred warriors, rapidly outstepped the clumsy, bulky European craft. These catamarans or multihulls had not only conquered the entire Pacific a millennium before Columbus conquered the Atlantic, but were more maneuverable, faster, and more seaworthy than any European craft. Yet there never was an attempt to capture or copy this technology by the Europeans. (Nor, conversely, did the Islanders feel tempted to copy the European craft.) Only today in the times of now over thirty-knot sailing multihulls for transoceanic racing have Westerners perfected the ancient designs. Here was a technology not transferred in spite of its efficiency and functionality.

Similarly, in the twentieth century, there exists a simple technology which would probably do more to help peoples escape poverty and drudgery than any single technology, but it is not adapted and is more ignored than rejected. I refer, of course, to birth control technologies which remain contrary to certain cultural values in spite of their obvious advantage for creating conditions for material wish fulfillment often equally desired by the very rejectors of the technology. Not all technologies are adapted, even when they have obvious advantages. And the resistance is a cultural one, not a difference between technological levels.

I have now examined a very few simple examples of technology transfers with their cultural implications. Let us now step back and note a few of the structural features embedded in such transfers.

1. In a transfer, an artifact is transferred. But in its original setting the artifact is paired with a human praxis, a technology is a human-technology relation. What is perceived as *useful,* in the typical transfer, must therefore make contact with a recognizable praxis, the familiar. Steel axes and knives are immediately incorporable. Here would be an example of a technology transferred between recognizable praxes.

2. But transfers may also occur between two entirely different praxes, with again the artifact being the focal object in the transfer. The use of sardine cans as shiny headpiece centers removes the artifact entirely from its previous field and places it in a new praxis. Here, what determines the new use relates to the already established praxes of the recipient culture. One could illustrate this just as well for early European culture. As Lynn White has shown, technologies such as "automated prayer wheels" became, once transferred, sources of power, the windmills of early medieval Europe.

3. Similar praxes may utilize different technologies and remain parallel. The resistance to adapt multihulls by Westerners and the coun-

terpart resistance to monohulls by Pacific Islanders illustrate this phenomenon. Only by changing the context, the cultural trajectory itself, would adaptation occur.

What these variants show—at the social-cultural level—is that just as at the individual level technologies must be understood as human-technology pairings, so must they be conceived at the social-cultural level. Only by understanding such gestalts will we be able to both understand and perhaps better prepare for transfer.

I want to conclude with an imaginative, rather than empirical example to illustrate the often radically different cultural gestalts which nevertheless can find focus in some fascinating technological artifact. I begin with the fascination of the early modern era, from the time of Descartes through even the present. This people is fascinated with automata. Their philosophical puzzles are filled with questions about how we might or might not be fooled by a cleverly contrived automaton. Is that store dummy "real" or a clever imitation of a person? In fact, I recently twice saw—once in France, again in Germany—a bit of theater in which, amongst automated dummies, there was a live human imitating the automatons in an inversion of the fascination. The human movements had become "mechanical" such that one had to look twice to detect the contrivance.

Yet this same people would walk past odd assortments of boxes which would emit human voices and music (speakers and amplifiers) without showing the slightest bit of curiosity. To have a human voice come from a most inhuman artifact was *not* fascinating.

Contrarily, the New Guinean highlanders were, on first hearing a 1930s Victrola, virtually astonished at the voice in the box—and indeed, some tried to find the "man inside." Would they have been fascinated to have a masked dancer, similar to one of the shamen in their own festivals but which was in fact an automaton, speak? I suspect such a phenomenon would have been far less fascinating than the voice in the box.

While the secret to these virtually inverted sets of fascination-to-familiarity lies in the equal/different praxes of the two peoples, there is hidden here a deeper clue to the cultural perceptions which both allow and resist the transfers of which I have been speaking.

I have been suggesting that in the complex issue of technology transfer, not only is more involved than the transfer of some artifact, but that entire cultural gestalts are involved. Moreover, in such transfers, there is always an exchange which is itself multidimensional. Those dimensions which range from fashion to force have both points of resistance and of opening which are often and probably endemically unpredictable. This is in no way to deny that there are often assymetrical aspects to such exchanges—patterns of dominant/recessive, of dependency/exploitation—and these are obvious in certain cases. But at the same time, such exchanges are not simply reducible to either economic or political factors. Until we understand the deeper dynamics of such exchanges, we are unlikely to do more than polemicize about the role of variant technologies and the ways they are culturally embedded.

This is also to suggest that technologies in ensemble are probably more like cultures than like tools. And were we to think in these terms, the whole problematic of much contemporary thought concerning technology would have to be rephrased. For example, the popular question concerning whether or not we can control technology, if thought of as parallel to a culture, would suddenly appear to be a rather misdirected question. We are rarely concerned with whether we can control a culture (indeed, whenever that question is raised it itself appears with ominous overtones in the light of recent history), yet the question of directions within a culture are in fact crucial to our social and political life. By pairing technologies as human-technology primitives, it may be possible to shift both our understanding of technology transfer and to escape what appears to me too much misdirected questioning currently dominating the debates over such issues.

4

Image Technologies and Traditional Culture

At the forefront of much scientific, but also of communication development stands a wide variety of "image technologies." In the scientific domain many of these technologies are designed to portray *interiors* without destructive or invasive procedures, such as CAT scans, MRS imaging, and sonograms, or, to investigate extreme macro- or microstructures of natural entities such as computer-enhanced photography or electron microscopy.

Image technologies within science *return observation to a perceptual dimension* which remains one of the primary valued trajectories in a largely visualist emphasis within science. Moreover, such perceptual visualism has all the advantages of immediacy, pattern recognition, and gestalt quality which characterize the informed style of vision appropriate to scientific observation. Such developments, although far more complex and intricate, continue what Derek de Solla Price called the "artificial revelation" which Galileo's telescope began.[1]

Microscopes, telescopes, the X-ray imaging, and now the above noted contemporary examples belong to a long history of perception-enhancing technologies which *embody* scientific observation. They are an essential element in the *technological embodiment of science in its instrumentation.*[2]

Accompanying this technological trajectory, and continuing the Galilean sense of discovery, there remains a fascination with what is shown, particularly when it is novel or previously unknown. We retain a sense of awe with respect to the latest depictions of forms of galaxies—spiral, crab, etc.—or with our newly attained ability to image atoms.[3] And, as with Galileo, what is revealed within the aura of fascination tends to override what is often less clear, the dimension of what is *concealed* or reduced by the instrument-enhanced vision.

While modernity clearly sides with Galileo, no longer doubting the mountains of the moon, the rings of Saturn, or the satellites of the planets, the churchmen of his time did have a point in that early telescopy also did have almost equally obvious limitations. Indeed, the much later parallel in which there was strong resistance to microscopy in biology was motivated not so much by religious reservations as by the awareness that "reading" the "images" contained far too much ambiguity and subjectivity (until aniline dyers helped create stronger contrasts in the specimens—not unlike today's use of "false color").

One could almost say that with both the more primitive technologies of the telescope and microscope there was a possible gestalt evaluation related to the ratio of revealing/concealing within instrumentally embodied vision. Defenders could rightfully note that "something" was being revealed, and that the macro- or microreality so shown was never seen before and therefore the device was a new means of discovery, while detractors could emphasize the concealment factors which should have led to critical doubts about "what" such entities could be. Indeed, the scenes were often complicated by objects apparently "seen" which were only later discovered to be effects of the instrument itself—instrumental artifacts.[4] Yet, in the end, the defenders have prevailed and there are today means of critique which in the experimental contexts include elaborate means of differentiating instrumental artifacts from observed phenomena. In the context of large instrument experimentation, such processes may be exceedingly difficult to determine and arguments about where and what is seen may proceed for years.[5]

This same history and process, however, also belongs to a much more public and *cultural* domain. Image technologies belong as intricately to our contemporary ordinary lifeworld, and *embody it in ways not dissimilar to the technological embodiment of science with respect to cultural dimensions.* The obvious equivalents to science's instrumentarium of image technologies in cultural life are the cinema, television, the growing spectrum of perception-enhancing and reproducing devices such as camcorders, VCRs, cassette players, CDs, etc.

The implication, in large, is also clear: what are the lifeworld equivalents to revealment/concealment? to instrumental artifacts to trajectories of development? It is at this point that the central theses of this paper may be entered.

> Modern science, since the Renaissance, became technologically embodied in its instrumentarium. Its perceptions are often instrumentally transformed. In parallel fashion, contemporary culture in its now global communications context is increasingly embodied through its instrumentarium, again image technologies.
>
> One may argue that not only was the irreversibility of a technologically transformed vision in science a major factor in scientific "seeing," but that its effects were *acidic* to previous modes of unmediated perception. The whole realms of micro- and macroentities revealed instrumentally overcame the narrower realities of unmediated perceptions, so that now mediated perception is fully sedimented within the scientific community. Here, within contemporary culture, one may argue that image or communications technologies are similarly acidic to prior modes of traditional cultures.

This investigation is an inquiry into the ways in which "image technologies" are non-neutrally acidic to all traditional cultures.

I. A Phenomenology of Technological Imagery

The first step in establishing the non-neutrality of image technologies with respect to traditional cultures will be a brief phenomenology of some of the structural features of technologically transformed perceptions. However, phenomenologically speaking, an initial qualification must be noted with respect to the prevailing terminology regarding "images" and their relation to "representability."

In a sense, "image" and "representation" belong to an implicitly Platonistic tradition within Western philosophy. Their use often implies that there is some "real" foundation, of which there are layers of secondary or tertiary "imitations" or "representations" and around which one may then generate any number of "realist/anti-realist" arguments. Phenomenologically the very term "image" becomes questionable. The so-called image itself is a *phenomenon,* that which can be seen. But what is seen must be analyzed in terms of its own context and field, within which it presents itself—it has its own unique mode of positive presentation. And, as will be shown, this very factor is one of the most subtle in the phenomenology of "imagery" itself.

I shall begin with an analysis of a few rather larger features of image transformations by contrasting mediated with unmediated perceptions of objects. Return to the transformations implicit in the Galilean use of a primitive telescope: telescopic vision is *framed vision.* Seeing through a tube changes the field within which the object is seen. The surrounding presence of the tube *changes the context* of the object seen.

Historically, although it was a complex instrument of compound lenses, even seeing *through* Galileo's primitive telescope was difficult. It has been likened to seeing a distant target through two small keyholes a yard apart! But, once targeted, the moon becomes framed within the limits of the telescope's tube. The "image"-moon is removed from its ordinary field within the sky and is framed within a now limited instrumental "field."

This transformation contains much, much more as well. The entirety of "apparent" space-time is also transformed. Relative motion, which is an indicator of space-time, becomes magnified as much as the moon as object. In a hand-held telescope it is very hard to "fix" one's visual object and bodily movement—jerks, inability to follow the now enhanced movement of the moon through the sky, coordination of eye-hand motions—all intrude into a previously already learned and "natural" seeing.

Spatially the moon is magnified as well. But it is magnified in the new context of difficult-to-control bodily motion. Both body and moon are thus magnified. And both factors are part of the now technologically transformed observational context. One may note, reflectively, that part of what occurs in the transformation is an implicit dialectic between the previously already sedimented form of unmediated seeing, and the new, as yet to be fully appropriated form of mediated seeing.

Third, in the selective magnification/reduction of telescopic seeing, "apparent" distance is also transformed. The body-moon distance is "diminished"—relativistically it is equivalent to say that one's body is now "apparently" closer to the moon, or that the moon is "apparently" closer to the viewer. But, more subtly, note that the very use of "apparent" implies at least a dialectic between technologically mediated vision and ordinary vision, or, at most, the implication that there is a primacy to the ordinary position.

Time dimensions are also implied in this transformation. The magnification of minute motion in the mediated situation also magnifies instantaneous aspects of time not likely to be noted in ordinary contexts. What was developed later in the history of telescopy—motorized mechanisms which allowed telescopes to "follow" the earth-celestial body motions and thus "fix" the object for continuous "fixed" viewing—was an attempt to reduce this time magnification.

Although I shall not here trace out all the implications to be noted in this preliminary phenomenology, it is interesting to note that much of the trajectory of development of the telescope revolved around increasing the *clarity* of the "image" (and thus the "transparency" of the instrumental capacity) as well as *producing* a fixed and repeatable object as the aim of the observation. In principle, such a fixing is the equivalent of creating a *photograph.*

II. Photography

The celestial object "fixed" in repeatable time and realistically "represented" is simultaneously an idealized photograph and implicitly reveals one goal of an experimental value. In its actual history, photography had a long preparation, but once invented it also led to a number of interesting cultural transformations.

As early as the Renaissance, the *camera obscura* had become an object of fascination. It embodied a kind of linear perspective in its inverted images, and according to Lee Bailey, later even became a major metaphor for the modern subject in Descartes.[6] Photography, later, allowed the *fixing* of the same kind of image found in the *camera obscura.* Photography itself was a nineteenth-century technology with the development of the Daguerreotype in 1839. Its invention was an immediate success followed by runs on optical shops for lenses for cameras, by hysteria from painters, and by fascination with the photographic representations which were produced. It was an immediate popular cultural technology.[7]

Phenomenologically, photography does many things noted previously in the case of the Galilean telescope, but also with differences: (1) it presents its "image" *framed,* i.e., it abstracts out of the living contextual field a specially focused and selected and narrowed image field; (2) in this case, rather than magnifying the body-object viewed motion, it *freezes* or *fixes* apparent motion, making it "still"; and (3) its result is a striking *isomorphism* or apparent "realism" between its image and the thing imaged.

I shall not return here to the already interesting studies which show that both (1) and (2) are learned modes of seeing which, while already prepared for and thus more invisible in Western Euro-American cultures, were more obviously in need of sedimentation for non-Western peoples. Instead, I shall take note of a few salient features in the mode of "representation" which followed the development of photography.

Using historical representations as a kind of phenomenological variation, recall the early prephotographic representations of Amerindians by the first European artists in the Columbian era: although the subjects are depicted as nude or with scant garments and feathers, the physiognomies as drawn do not differ much from standard Europeans. Conversely, when Perry arrived in Japan as late as the nineteenth century, local Japanese artists portrayed the Americans as very Oriental in appearance, again apart from dress which did more accurately represent the Western uniform. Both these cross-cultural examples show us something about a kind of naive viewing which is very unphotographic.

Similarly, while the history of—particularly European—painting did illustrate a trajectory toward "realism," viz. Dutch painting in the seventeenth century, it remained highly constructed and could not be compared to the postphotographic "photo-realist" style of the present. Interestingly, the "realistic" paintings of Amerindians in the nineteenth century by Catlin and Remington are also postphotographic. *The photograph, more than merely representing, "teaches" a way of seeing.*

Its "realism," however, retains within itself a distinct difference between it and what it represents: the very fix which fixes space-time, often metaphorically called "dead," gives the permanence of the photograph its own kind of *irrealism.* Its immobility, however close to the Greek or even scientific ideal of a "fixed truth," remains only that which is "represented." Adding motion to the picture radically changes that variable—the "moving picture."

III. Cinema and its Clones

Motion pictures followed still photography by fifty years (1889, with "Fred Ott's Sneeze"). And by adding motion to the previously still or space-time fixed representation, it gave a new sense of lifelike realism. The famous "L'Arrive d'un train en gare" of 1895 showed from a low angle, close-up, a steam train so realistically approaching the early cinema audience that they would cry out or move. Cinema—and now television and other audio-video forms—has obviously gone even further than the realism of the early documentary. And one important effect is the lowering of the realism/irrealism distinction still more strongly apparent in the photograph.

I shall turn to two very recent events to heighten this effect. (1) The Vietnam War of the sixties brought a kind of "television realism" into virtually every household of the West. Then under the freer ranging rules covering journalists (still derived from World War II), the battlefield and even civilian casualties were nightly brought before a populace at large. One effect was clearly repugnance at war, particularly that kind of war. By the 1990 Gulf War the "rules" for journalists had been radically changed. The news was clearly much more preselected. "Smart bombs" were repeatedly shown flying accurately into ventilators or hitting only focused targets (implicitly indicating that indiscriminate bombing, civilian casualties, etc. were being minimized), night-time successful Patriot defenses against Scuds, etc.

The "real footage" carries with it an implied realism, even though it was the *same* footage which was repeatedly shown day after day. It is the implied realism of the "documentary" which carries this ombra and is thus taken as real. It has taken until 1992 for the critical responses to begin to show what was *concealed* through the apparent realism of the news-broadcast, i.e., actual numbers of military compared to civilian deaths, low percentages of successful Patriot intercepts, offensive events like the live burial of Iraqi soldiers, etc. But what was most fascinating was the simultaneous *immediate awareness by the viewers that the news was being "cooked" and yet its planned effect was accepted and even celebrated.* It was as if the "realism/irrealism" of television was itself the new "real thing."

The same realism/irrealism phenomenon (2) is currently occurring with respect to the 1992 U.S. Presidential election campaign. Morality debates now occur over depictions which occur in soap operas as though they were real events (single mothers giving birth, teenagers being initiated into sexual activity), even though the viewers are fully aware of the presumed difference between the representation and the represented. The *medium* contains in its realism/irrealism presentation an effect in which "image technology" mediated and unmediated ordinary experience again becomes dialectical. What could be missed here is the parallel to the previously noted history of science examples. Although in the early days both telescopes and microscopes fell under severe criticism as means of gathering knowledge, within time both were sedimented into ordinary scientific consciousness in such a way that they became taken for granted.

IV. Phenomenology Redux

In the examples heretofore, I have taken early optical technologies in science and somewhat anachronistically compared them to more contemporary cinematographic technologies (and their clones). But there are special, artifactual aspects to cinemalike image technologies which should now be noted as well.

1. Some effects which dialectically contrast between visualization in ordinary sight and image-mediated sight have long been known. Seeing through lenses (a) diminishes depth. The object field is "flattened." This is particularly dramatic with television in which, contrary to Cartesian or linear perspective space, the size of the figure in the background remains the "same" as the figure in the foreground (viz. the pitcher on the field compared to the batter in baseball). (b) Contrasts are often different, thus the need today to enhance contrast, or even to move to "false colors." (c) At the gestalt level, reproductions are never entirely isomorphic with the objects reproduced—even today's digital enhancements which create "hyperreal" images are noticeably different from the actual objects.

2. But in the case of cinematographic-like technologies, it is the space-time *constructive* aspects which often stand out. Time reversals, flashbacks, special effects, and above all, discontinuities (very pronounced in MTV presentations), have been raised to a high technical art. Ordinary space-time is here technologically deconstructed and reconstructed in a bricolage of imagine space-time.

While it may be initially obvious that all these phenomena go into the transformation of contemporary cultural perception, it may not yet be so obvious how such phenomena non-neutrally impact upon traditional cultures.

V. Revivals of Traditional Culture

If contemporaneity contains a multiplicity of strands, including what is often called the "postmodern" and the conflict between modernity and the traditional, then the revival of strong forms of traditional culture at present may seem to be quite anomalous. Three such movements may be noted for examples: (1) the strongest such traditionalist movement is the revival of Islamic fundamentalism (although other fundamentalisms are also waxing); (2) if Eastern European Communism was initially viewed as a "modernization" process, its collapse in the last decade produced in its wake a revival of ethnic and nationalist revivals that currently characterize much of the strife in Eastern Europe; and (3) most prominently in the U.S. may be noted a kind of "neo-Puritanism" which is broader than long indigenous forms of religious fundamentalism or political rightism.

All of these movements are virulent, prominent, and apparently growing. It thus might seem, with respect to my thesis, that contemporary image technologies with respect to traditional cultures are producing results contrary to the thesis of non-neutral acidity.

Or, could it be that there is a deeper parallel than meets the eye in the very reemergence of virulent traditionalism within the context of contemporary image technology? Were the thesis that parallels early modern science with its instrumentarium again followed, one could at least note that Galileo himself lived in the very context of the Inquisition which historically was a response to both the rise of Protestantism and the expansion of colonialism (and a contact with the new cultures) in the New World.

There are clear aspects of all three of the above noted revivals which do appear to be *reactions* to what is often conveyed through image technologies: (1) One often-noted factor in the collapse of Eastern European Communism was the frustration over the lack of consumer goods in those countries. Related to this frustration was the portrayal of Western European and American societies which were rich in consumer goods—often the contrast was conveyed by image technologies, particularly movies and television. (2) Islamic societies often appear caught between conflicting responses to modernity. Does one modernize moderately by adapting high technology, but retain some core of Islamic culture? Does one lead a double life in which one is thoroughly Islamic at home, but thoroughly Western abroad? Or, does one react to modernization by reviving the religious pride implied in Islamic fundamentalism? In the cases of the more conservative Islamic movements what is of particular interest in this context is the contrast between the overt willingness to adapt totally contemporary weaponry and seek modern forms of productivity (oil production technologies, for example) while simultaneously censoring as fully as possible the image media from magazines to television. (3) In the case of American neo-Puritanism the role of image media is clearly prominent. Public radio is under attack for its presumed "liberal" tendency; the already noted soap-opera morality debates point to neo-Puritanical pressures upon television; and, as part of the same phenomenon of reaction, the attempt to accumulate more and more of the media channels by religious—primarily evangelical or fundamentalist—entrepreneurs, all belong to a response to the impact of image technologies.

VI. Image Pluriculture

An indexical question to open the way to the display of image non-neutrality is one which asks what is "imaged," and how that which is imaged occurs, which stimulates the various revivals or reactions of traditional culture.

Clearly the contemporary context is one of *global communication networks*. The image technologies occur in different ways in almost all the cultural contexts of the world. In countries in which television is seen as a tool to "leapfrog" from tribal village life into the twenty-first century (India, for example), satellites beam programs into central village halls in which a single television set is watched by all. In more developed countries television plus many other communications technologies are decentralized into homes and offices.

What is shown? Often an intellectual's temptation is to judge content in terms of a conflict between high culture and *popular* culture. It has been observed that "Dallas," an American soap opera, was perhaps the most universal of all programs in the last decade. In 1984, when I last lived in France, one topic of conversation amongst media personnel was how they could create a successful indigenous counterpart to retain their own French flavor! One can easily see that defensive cultures immediately can raise the question of program selection, control, or censorship. During the height of the Bhutto era in Pakistan the single most burning issue with respect to public television was how to keep Western portrayals of women off the television. But it is by taking a virtual "postmodern" example from highly developed countries that one can best see what is being delivered.

What is being delivered is *"pluriculture,"* a unique image technological portrayal of multiple "Otherness." Any international news program portrays a spectrum of countries and cultures in a single broadcast (Iran, Iraq, the Sudan, various European and American, Japan, India, South Africa, etc.). And, to take a most extreme counterexample from popular culture, MTV presents the most fragmented set of images from a multicultural mix of musics, fashions, ethnic traditions, and human races. Both the news and MTV are daily pluricultural bricolages.

The esoteric, the novel, the pluricultural—perhaps ironically—is not too different from Galileo's moon mountains, planetary rings, and unknown satellites. A tribal definition in which one's own group could be the only true human beings becomes increasingly difficult to maintain in a context in which the Others are present in one's own living room.

But the Other can also be a threat. Implicit in pluriculture is a kind of bricolage relativism. One may pick and choose culture fragments, multiply choices, and in the process reflectively find one's own standards often provincial or arbitrary—certainly not any longer simply a priori obvious. The pluricultural presence, moreover, characterizes not only the news or MTV, but virtually many kinds of programming. In one sense, of course, it is an actual reflection of a global experience, albeit mediated through image technologies which are the instruments by which pluriculturality is initially and sometimes solely viewed.

The parallel with the Galilean telescope is close—no human so far has actually viewed Mars first hand, any more than most of us have lived amongst the Berber, but both viewings are now part of our technologically mediated experience.

If the content is often pluricultural, the *form* should not be overlooked. The capacity of image technologies to fragment, deconstruct, only to reconstruct in a series of often juxtaposed, nonlinear imagery, *reenforces* the bricolage quality of what is shown, whether it be the short news item or the fleeting image of Madonna in a pose on MTV.

Although the above is obviously selective, it is also indicative. The portrayals on both the news and MTV stand in stark contrast to more traditional depictions of narratives, logics, progressions which are recognizable now as a more archaic form of broadcasting. The "talking head" of the academic television lecturer is simply received by most as boring. The animated talk show is a step up, but it is superseded by the forms which display the irreal realism of televisual "being there" in the on-scene camera or edited video. It is time to get "behind the scene."

VII. Edited Reality

How is the evening news or the MTV video constituted? In the case of the newsroom, the editing is done from the spectrum of multiscreens which display, simultaneously, the multiplicity of events going on in the world (these already selected from a vast variety of sources). The technician-editor scans the compound screens and selects and moves from one to another, forming a discontinuous narrative of news fragments. The result is a "constructed" result, fragmented, but deliberately designed.

Similarly, the video is spliced and constructed from shots and takes, remnants from the cinema, united by the narrative of the single song being sung over the discontinuities of the images. The "reality" is a constructed, edited reality.

I wish to note two aspects in this construction by editing. (1) The content which is edited is already the diverse set of multiple phenomena. The multiple screens from which the news is constructed refer to an ever increasing "tree" of possibilities which may be shown. But (2) the viewing which edits is itself a kind of compound vision. It scans and selects from among the fragments, itself a fast-moving process which must fit the style of the medium. (In fact, academicians should actually recognize a similar process in the now dominant contemporary mode of composition! Most of us use word processors which also mimic the same factors. Some experienced editors have noted that not only are manuscripts longer now—due to the ease with which footnotes, rearrangements, and other composition may occur—but are often marked by stylistic features which indicate that "text bits" may have been moved around in the composition-editing process speeded by word processing.)

What has been described are factors related to image technologies in operation. Such technologies also display "instrumental inclinations": cinema-like image technologies could, of course, restrict themselves to reproducing simply speech, but the inclination is usually more actional. That which is dramatic, reducible to the shorter bit, novel, even esoteric is more likely to be watched. And even when what is displayed is mundane—such as a Charles Kuralt travelogue through heartland America—it is made to be displayed as curious or novel. Image technologies "anthropologize" the natives.

In the inclination, a very interesting parallel again appears between contemporary communications phenomena and the introduction of instrumentation into early science. What instrumental mediations do is let one see what was previously unseeable, particularly the micro- or macrophenomena which eventually became the daily fare of science frontiers. What is fascinating is not simply that which is seen in ordinary vision. A counterexample from the history of art is instructive: one could have gone around the environment and simply held up a frame filled with plate glass if the point were simply to technologically mediate what is seen—but it is the *transformation* which occurs through variational change, magnification for example, which makes the newly mediated vision interesting. In the cultural counterparts a similar "magnification" phenomenon occurs.

Intellectuals deplore, in the popular media, the extravagant displays of sex and violence which become the common fare. Interestingly, this selectivity is a side-effect of an image trajectory. It magnifies what is initially titillating, escalates it into a kind of image-mediated frontier for entertainment observation. The kind of sex and violence portrayed does not characterize most mundane life or ordinary perception—but neither do the micro- or macroentities of today's frontier science characterize most mundane life or ordinary perception. So, although in radically different contexts, the *magnification-reduction phenomenon* shows the same thing. And in both domains the transformation phenomenon is acidic to traditional forms of seeing. Admittedly, our recent fascination with "buckyballs" may seem very different from the extremities of *Terminator 2,* but both are magnifications beyond the mundane made possible through image technologies.

VIII. Life Imitates Art

Galileo, in effect, created a new "art" through his telescope, a new way of seeing. In science such a way of seeing became so thoroughly constituted that today one may say that the carpentered products of science—perhaps more, or at least as much "technologically constituted" as "socially constructed"—take both their shape and measurement in terms of science's technologies.

In the ordinary lifeworld, image technologies produce, deconstruct, and reconstruct in a real/irreal artform, ways of seeing on the cultural plane. The results, again not unlike the history of science, are mixed. What is unarguable, however, is that the image mediation has made present a kind of irreversible awareness of pluriculturality which challenges, not simply the groups which today strongly react, but all our cultural forms. The beginnings of a questioning of Eurocentrism are first responses to what will become a much stronger transformation of awareness so long as we maintain the mode of image mediation which now obtains.

5

Technoscience and Pluriculture

Insofar as *technoscience* is a cultural product, it is the most powerful such outcome of Euro-American developments. Its reach is global. No culture exists without some relation to this phenomenon. Its artifacts often serve as "cultural instruments" which serve as an *entre* for other aspects of Euro-American culture into the most exotic of other cultures.

This cultural power has caused pause among some—particularly European philosophers thinking about technologies—to see the rise of *technoculture* (sometimes called simply "Technology" with the capital "T") as a distinct threat to past values and traditional high Culture (with the capital "C"). But, almost without exception, these thinkers remain "Eurocentric" in their perspective on technoculture: (1) they see it as a phenomenon unique to Euro-American history and metaphysics; (2) they see its maturation as a kind of "internal destiny" arising from Euro-American values; and (3) insofar as technoculture is seen as threatening, it is perceived as a kind of negative destiny *internal* to precisely this Euro-American trajectory.[1]

In this essay I shall recast this narrative common to many Euro-American interpretations of technoculture, and show certain ways in which the very contemporary situation of technoscience is situated within a different set of stresses relating to the rise of *pluriculture,* a distinct cultural product of technoculture, although related to the often occluded, forgotten, or neglected role of *other cultures* in our own history.

Metaphorically, my narrative is one which—to the degree that it agrees with the dominant Euro-American narrative—sees the rise of technoscience as a distinctly modern development originating at a crucial juncture of the sixteenth and seventeenth centuries, and which in the subsequent four centuries of development reaches like a tide to the shores of all the globe's landmasses. But, even if this tide is "tidal wave" in proportion, what cannot be overlooked is that any tide reaching a steeply sloped shore is also accompanied by an *undercurrent,* just as whatever destruction an incoming tidal wave may wreak, the outgoing return of the wave is also part of that effect. Thus, I shall pair the rise of *technoscience* with the appearance of *pluriculture.*[2]

I. Technoscience

The dominant forms of narrative regarding the rise of modern science usually place it in the late sixteenth and early seventeenth centuries, associated with the Renaissance and differing in some form from its antecedents in the ancient science that arose among the Greek philosophers. And, surely by the time of the contemporaries, Galileo in Italy and Bacon in England, much of the shape of modern science is discernible.

What counts as (1) the route by which science is "reborn," and (2) the characteristics which make modern science different from ancient science, are both important to this reconstruction.

If we grant that the early nature speculations of the Greeks, in the Presocratic through the Democritean era, were the birth time of "science" in European history, then for the most part this early science was not yet *techno*science. The dominant narrative—sometimes anachronistically—reads this history as the time of the ascendancy of a *particular emphasis upon theory* which occurs in parallel ways in both Democritus at the end of the Presocratic development, and Plato. Both the materialistic atomism of Democritus and the preference for ideality in Plato did become key elements in the rise of modern science.

The usual distinction that is made between ancient and modern science, however, is in the development of *experimental* method in contrast to the more contemplative and speculative "theory" of the Greeks. But experiment is also the appearance of one of science's essential weddings to *technology.* Modern science is *essentially embodied in its instrumentation, i.e., its technologies.*[3]

This is not to say that the Greeks had no technologies, or that many of the actual *measurements* which were made were not technologically constituted. Greek estimates of distances between the earth and the sun and moon, the diameter of the earth, and other often roughly accurate measurements of this kind, were often made with crude instruments interpreted through fairly more shrewd geometrical deductions. But Democritus's atoms were only inferred and their effects speculatively hypothesized in terms of shapes and motions. Nor were instruments seen in any way as *essential* to this more thoroughly contemplative theory-prone speculation.

Similarly, classical Greece was not noted for its technological innovations. The very "excellence" of Greek *technē,* which is romanticized by Heidegger, often served as a retarding force with respect to technological development. The well-crafted shield, the perspectivally controlled temple on the mountain, the decorated jug, could all be said to have been *aesthetically controlled and restricted.*

Of course this equivalence of art and technological objects obtained primarily within elitist and aristocratic circles—but it is to these circles that the philosophers belonged. Both common use-objects and the economic support of a slave-based society remained outside considerations and below the level of operation of the intellectual milieu of early science.

If now, we leap centuries to the late sixteenth- and early seventeenth-century European Renaissance, we see a rebirth of interest in a whole spectrum of interests, including nature speculations, which situated much of early modern science. I shall here concentrate on several marked differences between this science and ancient science, with particular emphasis on the shapes which were to become hallmarks of *technoscience.*

1. First, modern science was to become a new way of *seeing,* and by seeing I mean largely a certain type of *visualism.* Leonardo da Vinci's early anticipations of such a science saw sight as the primary sense, but developed along what today could be called "engineering" modes of vision. He invented the "exploded diagram," which equally and without any difference *represents* everything from human anatomy to his imaginary machines. His drawings of a fetus in the womb parallel the same exploded diagram approach as those of his visualizations of Roger Bacon's much earlier imaginary machines.

2. This seeing, however, is embodied in *specific technologies.* It is not "eyeball" seeing as such. Renaissance perspective drawing utilized graphed window or wire screened devices; variants on the *camera obscura*

were also used; and—of particular importance to early modern science—*optical devices* were employed in constructing vision.[4] Da Vinci's visualist science is simultaneously *technologically embodied* science. And, by the time of Galileo's refinement of the compound lens telescope, instrumentation is an essential part of "experimental" method. Science is first *technologically embodied in instrumentation.* Bacon, of course, was the first to much more explicitly celebrate this. He differentiated some twenty-seven different ways instrumentation could transform experience and thus make a "powerful knowing" which was also different from any merely contemplative knowing of the ancients.

3. If today we recognize how thoroughly the rise of modern science is associated with its technologies in the instrumental sense, we may fail to notice two other very contemporary aspects to this technologically embodied science.

Galileo, like Bacon, celebrated his instrument—the telescope—which created for him an "artificial revelation" of the heavens more powerful than anything either the ancients or the church fathers could bolster. It was through the telescope that he discovered what no deduction could show: mountains on the moon, satellites of Jupiter, rings of Saturn. But he did not leave it simply at that—he created the first "science public relations" magazine which was to publicize his discoveries, *The Heavenly Messenger.* Galileo's science was already a media-directed science.[5]

4. And, more, both da Vinci and Galileo found themselves in a second wedding to an aspect of technoculture. To have a science which needed technologies—even instruments—meant that these had to be *funded.* For this they turned to the princes of the time: da Vinci offered his services to the Duke of Milan, Il Moro; and, later, Galileo offered his services to the Medicis. In both their letters to these heads of the Renaissance "military-industrial establishment" of the day, *their enticement for the prince was the offer to invent and develop marvelous instruments of war!* Werner Heisenberg only repeated this history in a proposal which could have almost been copied from his predecessors, when he proposed to the war office of the Third Reich that his atomic science could produce nuclear powered submarines or an atomic bomb![6]

If technoscience is the science that is embodied in instruments and whose infrastructure is "military-industrially" supported, then even early modern science was already implicitly technoscientific. And to this extent and degree, modern science is differentiated from ancient science. But one more dimension is needed to constitute *contemporary* technoscience—and that is its *corporate structure.*

The making of a corporate technoscience occurs later—some of its origins may be traced to the eighteenth-century German chemical industry. Others arise in the developing university structures with "departments" and professionalization—again often associated with the German university which, in turn, became a model for the American public university in the nineteenth century. But the corporate solidification does not become fully developed until there is both a close tie between universities and government and industry, and until research becomes conducted with "teams" and in large laboratories.

Such corporate technoscience may be seen as a third stage in our brief history of science from ancient to modern to now, contemporary science. But it is now clearly sedimented as illustrated in the following indicators.

1. Big Science has now become the model for science research and development. The multibillion dollar, single instrument is now common: super colliders and the Hubble telescope are but two examples.

2. Most research is corporate team research, resulting in virtually all publications being jointly authored. The *New England Journal of Medicine,* one of the oldest medical science journals in the U.S., indicates that at the turn of the century 95 percent of all articles were singly authored, with today 95 percent being jointly authored. *Science* magazine recently indicated that the record number of signatures was in a mathematics article with some 194 authors!

3. And although the momentum of physics, with its early modern associates of astronomy and nuclear physics continues, both chemistry—which has almost as long as physics had billion dollar projects—and biology, which has its first billion dollar plus project (the Human Genome Project), have come into the Big Science corporate gambit.

This development now brings technoscience into its contemporary form. I have traced this large-scale history for several reasons, but one aspect of it is particularly important for what is to follow. Contemporary technoscience in its now technological and corporate form *requires a very complex, large, and society-supported infrastructure for the production of its "products."*

This, in turn, means that the autonomous production of technoscientific processes and products, even today, is restricted to those countries where this infrastructure has been developed: Europe, the U.S., a few of the major Euro-American derived enclaves such as Canada, Israel, South Africa, Australasia—and now two Asian countries, Japan and South Korea. Even the largest population countries—although close to autonomy in the cases of India and China—remain in some degree dependent upon the autonomous countries. In effect, of course, all countries are effected by technoscience and its reach is fully global.

II. Pluriculture

The reading of technoscience, to this point, might be said to be Eurocentric, in that however many modifications science has gone through to become contemporary technoscience, its history has appeared to be a linear one, running from the Greeks, through the Renaissance, to the Euro-American present. Such a reading is *distortive* and highly *concealing* of two factors which I shall now foreground simultaneously: cross-culturality and the history of technologies.

The first thing to note is that cultures *other than European* have played a role in the very development of technoscience from the beginning. For example, one of the foundational elements in science has been the use of mathematics as the primary means of both formalization and measurement. In Greek thought its source was largely Pythagorean—which in original form was Oriental and religiously embedded. Similarly, the rebirth of science in Renaissance times owed its very existence to Islamic culture. For although Plato may have had a more or less continuous existence in European thought, Aristotle and especially the science oriented Presocratics, were lost to the invasions of the very Northern Europeans who would eventually benefit from both the revivals of ancient thought and its often highly elaborated development—including the sophisticated development and use of instruments—by the Arabic scholars. Islam has frequently been the occluded Other in much of our history.

Then, regarding technology, what may be missed is the historical fact that periods of high theory development do not necessarily correspond to high innovative or technological development. For example, if the Hellenic and Roman periods in science and philosophy could be regarded as derivative periods regarding theory, this is not the case with respect to technological innovation. To the contrary, both Hellenic and Roman periods were highly prolific with respect to technological advance: most of the monuments of the "Wonders of the World" were built then; many of the "modern" lifemodes, including plumbing, indoor sanitation facilities, glassed windows, were invented then; construction methods including the arch, cement, roadbuilding, aqueduct water supplies, etc., were all developed in Hellenic-Roman times.[7]

Similarly, as Lynn White has shown, there was a virtual "technological power revolution" in medieval times which often borrowed from and adapted techniques learned in the early land explorations of the Orient and Far East, which led to the wind and water mills, the steel plow, the flying buttresses, large power machinery—albeit made of wood—of the shipping and building trades, etc., which stand in contradistinction to the logics and ontotheology of the medieval schools. Much of "modern" technology actually preceded, not followed, the birth of modern science.[8]

What this macroscale historical association shows is that innovative periods in the history of technology *correspond more to high periods of crosscultural trade and exchange than to high periods of philosophical or scientific theory.* (It would be tempting at this point to attack a now well-critiqued aspect of the dominant view of technoscience—that modern technology is largely the invention of and dependent upon modern theoretical science and that is what makes it different from premodern technology. I shall not yield to that temptation here, other than to remark that even contemporary very high technology, while being more closely tied to theoretical science than in earlier eras, has as only one of its important dimensions a tie to science.) Moreover, this association with technological innovation and cross-culturality hints at the very phenomenon I wish to forefront—the rise of *pluriculture.*

In terms of the narrative, there would be a balance if one could parallel distinctions between ancient and modern science, ancient and modern technologies, and between ancient and modern forms of cross-culturality. Were this to be the case, pluriculture would be the distinctly modern—even better, *postmodern*—form of cross-culturality.

If it exists as a unique form—as I contend it does—its appearance is closely tied to *technological mediation.* It may be located particularly within the matrix of the communications, information, and "image" technologies of contemporary technoscience.

All travelers and explorers, of course, have performed part of the pattern of tidal expansion and undercurrent. In their explorations they bring back "souvenirs," or exotic samples from the strange new lands. Early visitors to the New World returned with tobacco, potatoes, tomatoes, and corn, to cite but a few items, and these, it will surely be admitted, *radically and irreversibly transformed even European habits.*

Colonists, or those who live abroad for a much longer period of time than explorers or visitors, find themselves assimilating much more of the cross-cultural. If I stick to the largely culinary examples above, the colonist adapts not only to the material object, but its mode of presentation as a distinctive cuisine. Curries are as much a part of the English diet as pot roast; Indonesian food similarly is regularly part of the Dutch diet; and in New England America, Indian staples such as corn-on-the-cob, Indian pudding, and pot baked beans are whole-scale adaptations from indigenous culinary forms.

Each of these examples serve as indices for often forgotten forms of the already assimilated cross-cultural. Contemporary pluriculture, however, is more virulent and chaotic for several reasons.

1. What I am calling pluriculture is mediated by the communications, information and "image" technologies of the present. These technologies are (a) virtually simultaneous in real time, (b) two-way communicative, with (c) dominant "virtual reality" simulations of *what* is conveyed, and (d) globally networked.

Examples: some of the communications relating to any contemporary travel-conference setting demand communication speeds faster than even the six- to ten-day air letter—thus telephone and fax, both of which are also two-way communicative.

Televisual and cinematographic presentations convey a partial "virtual reality" sense which not only makes one feel one is "there," but which carry a kind of "realism" that is hard to deconstruct in any immediate audio-visual present.

Our reference world at present is global, with regular communications between the U.S., Europe, Australia, Asia, and South America. The network is highly multicultural.

2. If the mediation is as stated, the referential content is also different. If Cook's voyage to the South Pacific in the seventeen hundreds did, indeed, bring him into contact with a number of new and exotic cultures, these were, nevertheless, discontinuous, separated at least in the time needed to voyage from one island to the next, and still limited to a definite, geographical region of the globe.

Contemporary pluriculture is presented differently. It, first of all, is even commonly a prolific, virtually daily, multicultural presentation. Recently, the evening television news always contains at least (a) several Middle Eastern cultural presences; (b) both European and American references; and (c) since I am in Australasia as I write this essay, also both Australasian and southeast Asian items. To this list, we may add references to South Africa, and often to the Koreas, Japan, China, or India. Not even the most exalted medieval king had such a cross-cultural entourage in his daily presence!

Pluriculture, unlike older cross-cultural forms, is both more immediate and much more *plural.* I shall here introduce a symbol of this in what may be called the "compound screen" of contemporary vision. We are all familiar with the multiscreens of newsrooms or even department stores. In the newsroom case *each screen shows a different scene* and it is up to the technical editor to select and choose which image will be transmitted. Such vision, in principle no less technologically constructed than the user of a Renaissance perspective frame, is different nevertheless by virtue of its multiplicity and proliferation of its reference worlds.

3. This, in turn, gives technologically mediated pluriculture a unique set of characteristics which include (a) the already noted multiplicity of imagery, (b) its fragmentation into "bits" or "culture fragments," (c) a certain fluidity as they move into and out of presence, and thus (d) an overall "bricolage" character.

The question arises: can such a mediated and chaotic pluriculture presence take any societal shape? And the answer is that *it already has.* Its strongest presence is noticeable in the "advanced" countries, perhaps in a surface, but still a very actual way:

The Japanese businessman wears Western clothing both in the office and on world travels, but reverts to traditional clothing at home. Saki is still served, but alongside whiskey.

Hari Krishna dancers chant in every city.

Cuisines, fashions, religious fads, even intellectual trends are rapidly proliferated, adopted, then abandoned. We all "pick and choose" amongst the pluricultural fragments, and "edit" or construct our bricolage life culture. There is something quite "cinemalike" or even "MTV-like" about our cosmopolitan movements.[9] We are all in the process of "editing" or constructing our lives as in the newsroom image.

Now the question begins to take a different shape: is such a pluriculture still "Western" or "European"? It is clearly *not* Asian, *nor* African, but in a sense, it is also no longer simply European or even American. Its shape is different from all of these. Those who take these phenomena negatively, see in the *acidity* or non-neutrality made possible through these technologies the appearance of pluriculture as a "loss of a center," "the celebration of relativism," or the decline of Eurocentrism. Those who take the phenomena positively see value in multiculturalism, a new richness of choices, and the basis for a New Enlightenment which must reshape the cultural politics of tolerance and cosmopolitan life which once came about with respect to the eventual solutions to the religious wars of the post-Reformation.

III. Postmodernism and a "New Enlightenment"

In an important respect the route I have taken is merely another way of arriving at our intellectual present. I have done so by highlighting technoscience and culture, but the dilemmas of our era have also been felt and are reflected in our current philosophical debates.

The "later Heidegger" has been adopted by the "French School" and its American cousins. Here the "end of metaphysics," the deconstruction of foundationalism, the new relativisms and pragmatisms—all postmodern phenomena—have seen what I have identified as the nonneutral acidity of technoculture with respect to all traditional cultures. Philosophy has been shown to be, within this setting, highly culturally embedded, and as its culture changes it must—as it has had to do many times before—rethink itself. Today it must do so in the midst of, and with the explicit recognition of, *technology and culture.*

And if the still bricolage character of skipping through the culture fragments, particularly obvious in today's popular and youth cultures, strikes the traditional philosopher as superficial and not yet formed, then one must also see that the architectonics of our variants upon "systems" and "metaphysics" appear increasingly narrow and, while perhaps "deep," also highly parochial. More significantly, if what much of the turmoil of contemporary philosophy has pointed up is the *cultural and perspectival embeddedness* of philosophy, then the very shape of culture become pluricultural bespeaks a different bed from which to reflect. It is this deconstruction of our taken for granted traditions which the postmodernists have magnified for us. They have claimed that there are no "master narratives" and have substituted precisely the multiperspectival set of visions which are episodes and epistemes for foundations and master narratives.

But even if the critical thrust of postmodernist philosophies is correct—and I am deeply suspicious that it is—it does not point to any solutions or ways to address the new conflicts which take shape within a pluricultural world. Here, too, we have become newly aware of the cross-cultural in that our wars and social transformations have today taken on an increasingly cultural dimension. I would claim that the revival of the various "fundamentalisms," the resurgence of ethnicity in neonationalist form, the strong thrusts to smaller enclaves of quasiprovincial nature, are all also reactions to both the reality and the as yet unsettled response to the pluricultural.

And in the face of this situation I find myself equally empathetic to those who defend a neo-enlightenment set of social and political values. The enlightenment, after all, reached a domestic solution for the long period of post-Reformation religious wars which were an earlier form of mini-multicultures in conflict. But the solution was one which *changed the very nature of religions* by making them take any of their foundationalist and absolutist tendencies into a private or contained realm, in order to participate in a new and largely secular political order. Religious tolerance was a "price" for religions read in the pre-enlightenment forms.

In this analog, the pluripalate dilettante is at least a harbinger of a new enlightenment, insofar as he or she can see that there is no single "core" or "foundational" cuisine and that many can be equally tolerated, or better, deeply appreciated. There is a kind of cosmopolitanism implicit in this phenomenon which will always offend the countryman. But if, through the mediation of now omnipresent information, communication, and image technologies *there is no more "country,"* for the "global village" is not a village, but an urban complex of global diversity, including all the ethnic neighborhoods contained within the city, then neo-enlightenment cosmopolitanism may be the order of the day.

I shall not pretend to know what the end to this history may be. Our past is full of both pluralistic times of productive peace—such as the two century period of pluralist flowering that occurred in Spain with the tolerance of Christians, Moors, and Jews, which produced not only the resurgence of the very roots of modern science noted earlier, but also times of restrictive and oppressive monoculture, such as the Inquisition, which ended the earlier Spanish tolerance.

What is clear to me is that we already live in the volatility of the pluricultural, perhaps ironically brought to us and in its contemporary form, by our own highest cultural invention, technoscience. Both phenomena belong together as the binary stars of the contemporary.

PART 2

PRACTICING POST-PHENOMENOLOGY

This part opens with an intricate examination of human perception, "Perceptual Teleologies," which is that most explicit example of how a *variationally centered* practice of phenomenology operates. It is, thus, a crucial essay within the group. There I show both that bodily perception has a *structure,* but that very structure yields a polymorphy to perception. It is *multistable.*

In chapter 6, the background points to the postmodernist context in which so much Euro-American philosophy finds himself with a concern over the "text" or its correlates. It is this which I seek to counter. In chapters 7 and 8, which in some ways are more standard pieces of "continental" criticism, I turn first to Paul Ricoeur, and then to Martin Heidegger. In the first case it is again the phenomenon of *reading* that shows itself, and in the second a variation upon cultural macroperception.

The chapter "Feminism and the Philosophy of Science" is my first foray, in print, into a direct critique of feminist texts. It has long been my practice to read and teach, often for a few years, before committing to print any results. Thus, although today virtually all my classes in either the philosophy of science or of technology contain large doses of feminist critiques, this 1991 essay is a first in one of my collections.

What I hope the reader will gain from these originally separate and distinct essays, is the sense that postphenomenology is the practice of both a perceptual and cultural *variational theory.* Perspective, if limited to one or even one dominant take upon the world, would be limited indeed. If all perceptions are "profiles" as Husserl called them, then depth and adequacy only can come by a more radical set of multiple and even refractive perspectives.

Although at first it might seem strange, that is what is common to both the focus upon technologies and this most recent attempt to sympathetically read feminist critiques. One learns as much about "eye-ball" perceptions through the variation into instrumentally mediated perceptions as, hopefully, males might learn by trying to see the world through the experiences of women.

The same approach was also behind my discussion of both the compound seeing which image technologies invite, and the inevitable confrontation and accommodation to the plurality of cultures which characterize our times. It is my contention that the variational approach of postphenomenology provides a means of both respecting and yet seeing a deeper multistable structure to these possibilities.

6

Perceptual Teleologies

I. A History of Perception

I begin with Derrida's enigmatic claim, "there never was perception."[1] Were one to look at the drift line of much contemporary Euro-American philosophy today, the emphasis might seem concentrated upon "narrative," "reading," "discourse," "texts," "textuality," and the like, seemingly linguistic, or even more narrowly, writing phenomena. And it might seem that the *perceptualism* of earlier phenomenology has either been forgotten or suppressed. Here I shall argue that there is, contrarily, *always* perception, and that its teleology may be detected within the very phenomena which surround today's revival of the "book metaphor."

I suggest that the "book metaphor" is itself a return to an older and essentially *prescientific* tradition, whose roots are to be found in the medieval and early Renaissance traditions. Could it be that this drift of Euro-American philosophy, not unlike its cousins in analytic essentialism (of the foundationalism sort pronounced terminated by Richard Rorty), who have discovered in medieval logic a new interest, is in the continental context equally a revival of the primacy of reading and writing?

I suggest this from an insight provided by Michel Foucault, who will, as a background figure, here serve as a pivotal point between what I shall call phenomenological perceptualism and poststructuralist textualism. Foucault noted, in *The Order of Things,* that

> The great metaphor of the book that one opens, that one pores over and reads in order to know nature, is merely the re-

> verse and visible side of another transference, and a much deeper one, which forces language to reside in the world, among the plants, the herbs, the stones, and the animals.[2]

May I point out that this is the era which is historically prior to the scientific era, the latter era in which one might say—Foucaultlike—that perception was "invented." This book metaphor, just at this formative juncture, became even stronger before disappearing the later age of science.

> In any case, such an interweaving of language and things, in a space common to both, presupposes an absolute privilege on the part of writing.
>
> This privilege dominated the entire Renaissance, and was no doubt one of the great events in Western culture. Printing, the arrival in Europe of Oriental manuscripts, the appearance of a literature no longer created for the voice or performance and therefore not governed by them, the precedence given to the interpretation of religious texts over the tradition and magisterium of the Church—all these things bear witness without its being possible to indicate causes and effects, to the fundamental place accorded in the West to Writing. Henceforth, it is the primal nature of language to be written. The sounds made by voices provide no more than a transitory and precarious translation of it. What God introduced into the world was written words.[3]

(Shades here of Derrida's stance vis-à-vis Husserl in *Speech and Phenomena!* Save that it was written a year prior to that text and refers to the much more ancient juncture between pre- and scientific eras.) Today's antiperceptualism and antiphenomenology thus finds its ancient echo at precisely the juncture of the book metaphor. It will be my task, similarly, to "reinvent" perception, particularly with a concern for its hidden role within the text-world which today constitutes the surface of the current tidal drift.

The tradition, the *phenomenological* tradition from which I shall draw, I shall take to be a broad one. And I wish to undertake this introductory location of the role of perception within that tradition by distinguishing between what I shall call *microperception,* whose emphasis is upon bodily-sensory dimensions, and *macroperception,* which emphasizes cultural-hermeneutic dimensions. The first emphases are easily associated with the blatant perceptualism of Husserl and Merleau-Ponty, and the second with the more implicit cultural perception found in Heidegger and—for reasons to be seen—Foucault.

Husserl, the grandfather of phenomenology, is well known for his perceptualism, and I need not rehearse the theory in full detail here, but will note several salient features: (1) Perception is the base and primordial originary mode around which most other experiences revolve, with perception its own peculiar phenomenological essence.[4] (2) The form perception takes is that of a gestalt, or figure against the ground, with every object situated within a surrounding context.[5] (3) And every such presentation is positional—presented *to* a perceiver in a particular position with the perceived aspect a "profile." (4) The presentation itself is always multidimensioned and complex, with a phenomenological denial of "primary" and "secondary" qualities to the experience. And (5) there is a "depth" to all such perceptions with both manifest and meant characteristics—objects present profiles, but profiles carry with them the significance, "having a backside," etc.

Here is microperception: sensory, fundamental, and reflexively referring to bodily position. It is the primitive, phenomenological model of perception.

Merleau-Ponty, particularly the early Merleau-Ponty, radicalizes the perceptualism of Husserl. "The perceived world is the always presupposed foundation of all rationality, all value, all existence."[6] He accepts the five characteristics noted in Husserl, but emphasizes even more dramatically the position of the incarnate body in its motility and kinesthetic interaction with the environment. And, as with Husserl, the primary focus is on what I am calling the microperceptual, i.e., sensory and bodily perception. The chapter titles of the *Phenomenology of Perception* are indicative enough: "The Spatiality of One's Own Body and Motility," "The Synthesis of One's Own Body," etc. Insofar as this focus characterizes most of Merleau-Ponty's work, it remains within the Husserlian microperceptual trajectory.

There is, however, one feature which occurs in Merleau-Ponty which takes his analysis beyond the Husserlian context. In addition to perception being multidimensioned, it becomes ambiguous or *polymorphic.* The recognition of such polymorphy occurs as early as "The Primacy of Perception" and is there associated with an aesthetic phenomenon: "It is a remarkable fact that the uninstructed have no awareness of perspective and that it took a long time and much reflection for men to become aware of the perspectival deformation of objects."[7] It is here that an opening to what I am calling the macroperceptual already occurs.

This opening to the macroperceptual is made even more explicit in *The Visible and the Invisible:* "I say that the Renaissance perspective is a cultural fact, that perception itself is polymorphic and that if it became Euclidian, this is because it allows itself to be oriented by the system. . . . What I maintain is that: there is an informing of perception by culture which enables us to say that culture is perceived."[8] Here we have macroperception.

However, if we turn to Heidegger, we note something different happening. First, with the exception of the *Die Grundeprobleme der Phänomenologie,* there is not much explicit discussion of perception, and in *Being and Time* we find perception either dismissed or absorbed into the existentialia of understanding. "The kind of dealing which is closest to us is . . . not a bare perceptual cognition, but rather that kind of concern which manipulates things and puts them to use; and this has its own kind of 'knowledge.' "[9] Not that microperceptual analysis is lacking. The very concrete discussions of "the spatiality of Being-in-the-World" of Dasein are clearly descriptions of bodily existence. But spatiality takes its shape in a much broader macroperceptual context. And the macroperceptual is what *contexts* the microperceptual. Perception is taken into understanding and interpretation, or to rephrase, into what may be called a historico-cultural or "hermeneutic" perceiving. And the further into the "later" Heidegger one moves, the more prominent does macroperception become, until the later version of the phenomenological gestalt emerges as the historico-cultural emergence of the enigmatic fourfold.

Now, finally, Foucault: It is clearly not the case that perception is forgotten. Foucault undertakes "archaeologies" of perception. Indeed, the radical shifts of epistemes through which the clinic is born, in which "man" is invented, by which the punishment by the sovereign is first worked out upon the body of the condemned in public so the victim's body experiences the "body politic," and then becomes transformed into the invisible punishment which happens in the Gulag, are all "histories of perception." And even microperception undergoes shifts, from the history of the historico-perceptual descriptions of medieval "botany," to the extended tables of colorless (extended shapes only) of the Cartesian era, to the microstructures of contemporary biology, perception itself seems to be transformed. At the least, what had been the apparent stability of Husserlian microperception, becomes, with Foucault, the hermeneutic ambiguity of refractive epistemes.

One might be tempted to read this capsule history of perception as one that juxtaposes the microperceptual (sensory-bodily) dimension with the macroperceptual (cultural-hermeneutic) dimension. Were that to be the case, Hurserl and Merleau-Ponty would stand on one side, Heidegger and Foucault on the other. But such a juxtaposition is only apparent, and I should like to suggest that there is a way to unify this history from its own deepest insights. That suggestion takes the form of noting that there is an interrelationality between the micro- and macroperceptual dimensions not unlike that gestalt notion that originated the phenomenological theory of perception.

The matter may be put simply: there is no bare or isolated microperception except in its field of a hermeneutic or macroperceptual surrounding; nor may macroperception have any focus without its fulfillment in microperceptual (bodily-sensory) experience. Yet in the interrelation of micro- and macrodimensions of perception, there may lie hidden precisely the polymorphic ambiguities which most particularly emerge in the later work of Merleau-Ponty and Foucault in particular.

It is at this now contemporary juncture that I wish to enter the situation, and in the analyses to follow, the explorations will be precisely of those polymorphic ambiguities. I shall be suggesting that (1) the ambiguities are not simply polymorphic, but are *structured multistabilities;* (2) by following now an expanded phenomenology of perception which links micro- and macrodimensions, clues may be discovered as to the shapes of these structures; and (3) what will emerge is the recognition that without an enlarged perceptual analysis an entire existential domain may be overlooked.

In what follows, it will be seen that I am following standard phenomenological practices beginning with object or noematic domains, and through the rigorous use of *variations,* ascending toward the juncture of micro- and macroperceptual interstices.

II. "Cubes," or a Multidimensional Miniworld

Suppose a visual miniworld of drawings (representations) and objects which may be called a "cube world." My first example is the drawing of the now traditional and familiar Necker cube (see figure 6.1). Everyone is familiar with its apparent bistability. Either one sees the cube as tilted "upwards" toward the viewer, or, in a reversal, tilted "downwards." These two visual profile variants I shall term 3-d^f and 3-d^r. Note that corresponding to the variants there is also a detectable, if subtle, shift in the apparent position of the viewer, with the first placing my embodied

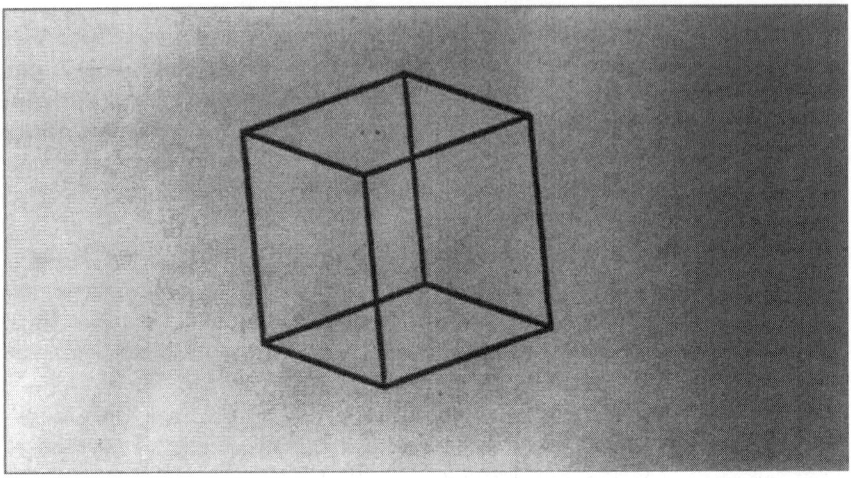

FIGURE 6.1

position somewhat "below" the central plane, and the second somewhat "above." Here will be one constant clue in all the analyses to follow, i.e., the phenomenological correlation of noema and noesis with respect to the thing seen and position from which it is seen.

Now, as those who are familiar with *Experimental Phenomenology* will already know, the conventional profile variants are not the only ones which the drawing may be said to "represent."[10] There is a third stability, a two-dimensional one, which may also be gestalted. I previously have suggested this by redescribing the figure thusly: the central parallelogram is the body of an insect; the hexagonal outline is a hoe; and the lines from the parallelogram to the hexagon are the legs of the insect, and "voilà," profile 2-d. Now note here that the redescription is an instantiation at this simple level of micro- and macroperception, in this case with the hermeneutic tale as the vehicle of letting the "insect" be seen. It is a Heideggerian-Foucaultian strategy. The actual micro- or bodily-sensory fulfillment "follows" the linguistic-hermeneutic tale. It is a mini-episteme or discourse-perception.

But we need not rest here. Perceptually there is more to be found. Again, as previously demonstrated in *Experimental Phenomenology,* there are more eidetic perceptual possibilities, which could be demonstrated with actually constructed three-dimensional objects (which the drawing could be taken to "represent"). This is shown in figure 6.2. You can immediately see that the previous insect now may stretch itself and constitute a new three-dimensionality and its reversal, thus establishing the series 3-d^f, 3-d^r, 2-d, 3-d^f, 3-d^r. This eidetic series, again as demonstrated

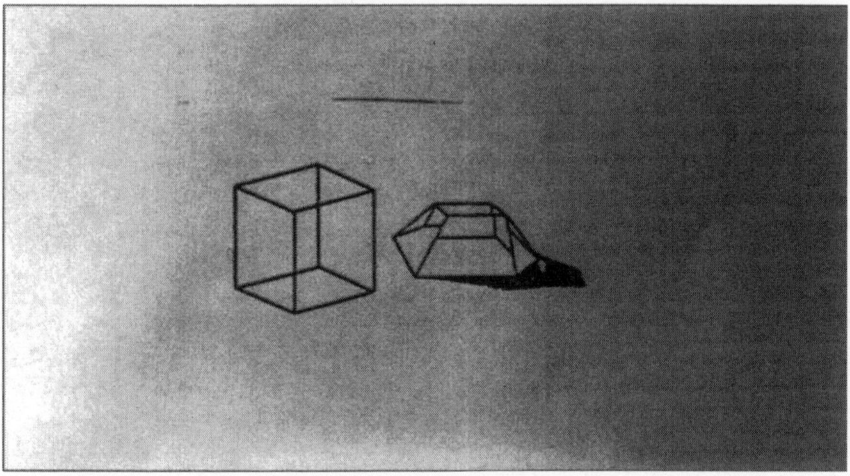

FIGURE 6.2

previously, belongs to the entire series of presumed Necker cubes (see figure 6.3).

This is to minimally suggest a trajectory of possible (eidetic) perceptual variants. Although several have been demonstrated here, there is an indefinitely large such set, which not only may be shown, but constructed in three-dimensional models that go far beyond the conventional analyses. Here is a simultaneously "phenomenological" and "deconstructive" strategy which shows something about the richness of such objects and representations. What emerges is a kind of set of "mini-epistemes" which occur in this interrelation of micro- and macroperception.

One could also ascend to a higher level of complexity and show the same thing. Instead of line drawings to "represent" the possible perceptions, a shaded version of a Vaserely-like cube drawing with its greater sense of depth and three-dimensionality is also indicative of the same progression (see figure 6.4). Again, I may speed the process of phenomenological seeing by associating the representation with its possible objects. The figure could be of a building with an inverted balcony; or, it could be of an overhanging structure with the small cube "hanging" up in its corner. *Or,* if we wished a high-tech hermeneutic interpretation, it could be of a cubical "space ship" about to dock with the now fully cubical "building" still in the distance.

You should begin to see that the same progression of indefinitely (though not infinite) profile possibilities belong to our now shading-enriched miniworld of Vaserely-like drawings and objects.

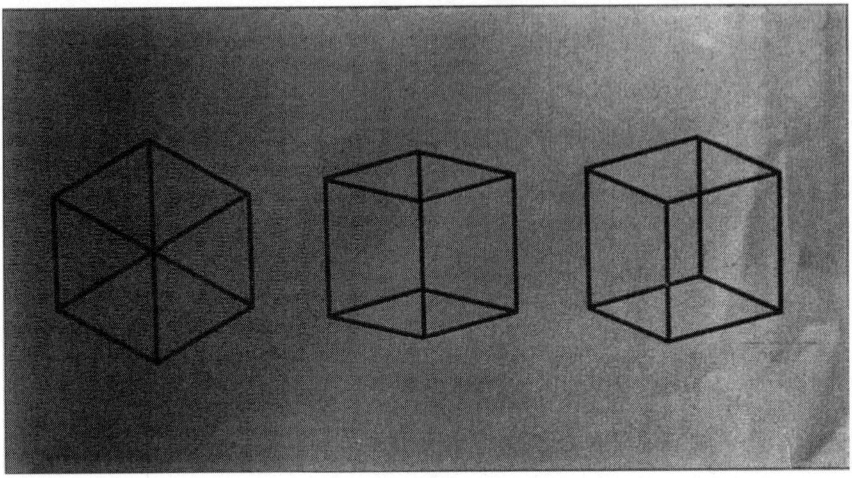

FIGURE 6.3

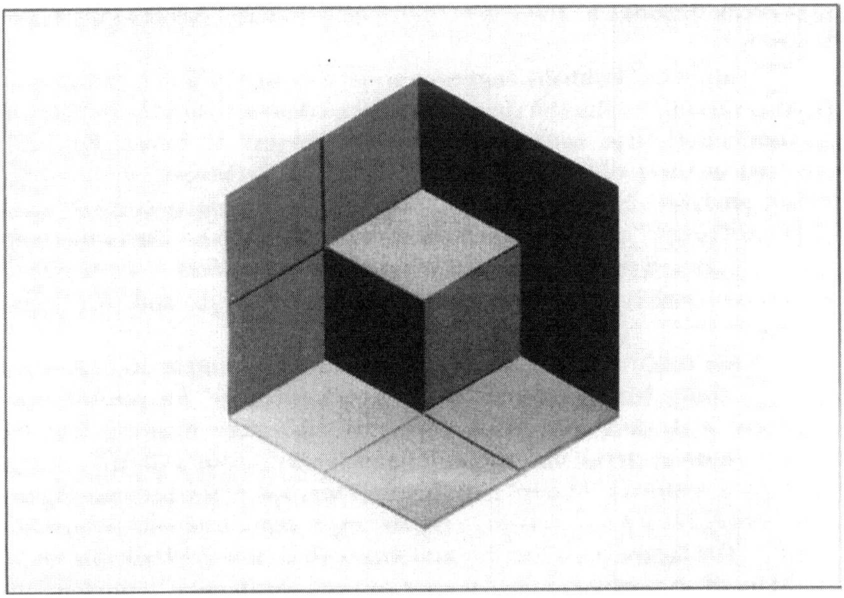

FIGURE 6.4

Once again note what I have been doing. First, I have been applying straightforward phenomenological variational practice to this set of representations and objects. It is an example of eidetic variations, to be sure, pushed more concretely and specifically than much of the literature illustrates, but nevertheless in a quite orthodox fashion. But, second, by relating each profile variant with a story (and with imagination, you should be able to see that the story variants *could* be as polymorphic as the strictly perceptual referents), I have activated the relation between what I am terming the micro- and macroperceptual domains. Were I to use the Merleau-Pontian phrase, I am demonstrating that "culture is perceived." But in this interrelation of micro- and macroperceptual dimensions, there is a rich range of multistable possibilities and thus of possible teleologies.

III. "Positions," or "World as View"

As we approach the *perceptual* role implied in reading, I shall turn to a different example set, here drawing upon historical and cultural variants. I am concerned here with the polymorphy, not so much between representation and represented, but between presupposed and implied *positions* of viewers in relation to the scene.

For example, let us take a series of depictions of gardens from two different periods and cultures (see figures 6.5 and 6.6). Note, now, several things. (1) The perspective *from which* each of these garden depictions is taken is either an *elevated* or an *overhead* one. What varies is simply the degree of close-up to greater distance, or the degree of height from, as it were, "stepladder" to "bird's-eye" height, or, in the case of the overhead, a directly overhead "bird's-eye." (2) Note, just in passing

FIGURE 6.5

FIGURE 6.6

first, that the *actual* position of viewing is what I shall call "*reading position.*" First to be noted is one's actual position, which were you in your study, you would tend to be looking somewhat downward, as when reading a book. (3) But in addition to the perspective and the literal or actual reading position, there is what I shall call the "presumed position." That is, of course, the implied position *from* the reflexive reference of the perspective. That is, "presumed position" is where you would be were you to be on the "stepladder" or "bird's position." (We noted this briefly before, in the sleight shift in presumed position in viewing cubes.) (4) Note, finally, that actual position and presumed position need not be identified, although there will be some polymorphous relation between the two.

All of this is phenomenologically microperceptual. It is a taking account of the bodily-sensory dimensions of seeing as ——— which occur in our garden depiction context. Let us ascend, briefly, to a more historical-anthropological dimension which indicates something of the macroperceptual. (1) First, the perspective with its implied presumed position of viewing in elevated or overhead positioning is clearly *cross-cultural* and historically diverse. Nor do the perspectives and presumed positions need to have been structured by such specific inventions as "Renaissance geometrical perspective." Such a perspective is simply a specific variation upon the noema-noesis of depiction. (2) Note, however, that each of the examples come from what we would usually call "civilizations" or, more specifically, *literate cultures.* There is some form of writing, hence of reading, in each of the examples so far. (3) The attention I am calling to this association may relate, more subtly, to the two positions, the actual or reading position and the presumed or reflexively implied position, from which the perspective is taken.

Now, take a look at two examples of depictive artifacts from nonliterate cultures (see figures 6.7 and 6.8). In figure 6.7 (Australian aboriginal), each individual bird is depicted, *not* from elevated or overhead perspective, but from what might be called horizontal or "standing position." The bird is viewed from a "lower altitude." But, secondly, there is a kind of atomization to the composition. Each individual bird implies a separate presumptive position. Here the presumptive position, compared to the actual viewing position, displays different relationality, a *diffracted* relationality.

Figure 6.8 (Andean tribal example) is even more extreme. Each individual figure (human, canoe, animal, etc.) by itself is depicted from the same horizontal or standing position, but the composition as a whole does not display an organized unity or synthesis of such perspectives. Such depictive phenomena are common in many nonliterate cultures. And they are clearly starkly different from the previous set of synthesized, elevated or overhead perspectives.

Now, in the final example in the "world as view" series, I turn to an example from ancient Egypt (see figure 6.9). Here is a curious amalgam of what we have seen in the two previous variations. First, note that

FIGURE 6.7

FIGURE 6.8

there are two "inconsistent" perspectives within this depiction. The workers and the vines are, while doubled, both seen from the horizontal or sideways position. The doubling, one position above the other, displays vestigially the same discontinuity as in the nonliterate set, but comes closer to a synthesis in that they are at least parallel.

But the water pool is more radically discontinuous, in that it is depicted from the overhead perspective and combined with the two just mentioned horizontal perspective/positions. Yet the entire composition does have a doubled (overhead plus horizontal) quasi-unity.

Now note something curious about the historico-cultural context of the Egyptian example—it occurs at the point of one of the inventions of *writing*. And this writing displays two roots. The first is that of pictographs. Pictographic "writing," if it can be so called, often displays certain similarities to the nonliterate set I have shown, although usually displayed in some set of series of pictographs. The second root of Egyptian writing is quasiphonetic. For even if the "sound" is depicted as a figure, the reading and deciphering process which interprets it takes it as phonetic. Thus we have a *doubled* or diffractive process within the act of reading itself.

I suggest that something isomorphic with the diffractive reading process is occurring with the Egyptian garden. The doubled perspective/position reflexively implied by the depiction would be already familiar to one whose reading praxis is of the type described. Here lies a secret link between reading and perception in both its micro- and macrodimensions.

FIGURE 6.9

IV. Perception and Praxis

There is a generalization that can be made concerning the previous examples—although not an entirely strict one. Both of the elevated perspectives (figures 6.5 and 6.6), from a medieval and a Chinese garden, came from *literate* cultures. Such depictions were not only "realistic," but taken from elevated or "readinglike" apparent positions.

Is there a deeply embedded *praxis* that permeates what could be called a privileged perspective within literate cultures, which differs from praxes in nonliterate cultures? I have argued that such is the case—for example, in navigational praxes, there are starkly different perspectives taken from European derived map reading and privileged overhead perspectives of "navigational position," compared to what may be called "embodiment" positions taken by many South Pacific navigators from Polynesian, Micronesian, and Puluwatean (nonliterate) traditions (see chapter 7 for a more detailed working out of these differences).

Does *reading* and its praxis perspective permeate, not only the development of the kind of representation noted in the above differences between Aboriginal and Andean examples, and those of the medieval and Chinese examples? The answer cannot be unqualified, since inland Aboriginals also depict waterway and sacred grounds as "maps," and the Nazka of the Andes are makers of large structures of lines and beasts which can actually best be made out from the air.

The exceptions indicate that, cross-culturally, each of these peoples has the ability to occupy imaginative bodily positions never actually taken. But, on the other hand, the sedimented and habituated taking of a privileged position *inclines* toward certain, rather than other, depictions. I am suggesting that there is, at a very crucial period of the formative period which gave rise to the "modern," such a sedimentation.

First, the production of written materials lies immediately before the Renaissance. Printing and the Western rediscovery of movable type; the move to a vernacular; the rapid dissemination of works throughout European universities—studies have shown that the dissemination of Copernicus's work rivaled that of contemporary times—all were part of a spreading of reading praxis. Leonardo da Vinci, the illegitimate son of a middle-class father, was noted for his large library (166 books—very small even for a graduate student of today).

Moreover, this rise of reading praxis was also accompanied by the increasing emphasis on the *visual* during this period, which many authors have noted (Lucien LeFebvre, Michel Foucault, Walter Ong, and, cf., my earlier *Listening and Voice).*

Within reading praxis, within a phenomenology of its actual mode of perception, lies precisely the "elevated perspective," the need of a fixed "text," the stability of that "text" as page, which gets reflected in the range of phenomena from depiction to navigational praxis to the preference for vision as a privileged sense.

With reading there *is always perception,* but a particularly structured perception. It is a perception which, normally, carries with it a dampening of bodily motion, a fixed place for its object, an enhancement of the visual, and the privileging of an elevated or overhead position. This praxis is never explicit, although it is implicitly taught and followed. It becomes a way of "being-in-the-world" particularly within literate cultures, those cultures which become "modern."

Today that praxis has become highly adumbrated by other forms of "reading" which are not—except metaphorically—strictly reading at all. Here I refer to the rise of "image technologies" which, while still *viewed* from a usually *fixed* position, now begin to vary the "text" with that which "moves" and which develops a virtual "movement" of bodily positionality (as in television, cinema, etc.)

This multiplicity of positionality—always present in some ways, but now enhanced by the multiple praxes of our different ways of "seeing"—is the emergence of the polymorphic seeing which begins to characterize the *postmodern.* Only the romantics will nostalgically see this as an age of the "decline" of reading. The postmodern will see it as a proliferation of ways of "reading" now become metaphorical in the image technologies of the present.

V. Conclusion

The trajectory I have followed is one which has attended to the perceptual, both in its micro- or bodily-sensory and its macro- or cultural-hermeneutic dimensions. But it is a trajectory which also demonstrates that there is *always* perception. Its variability, so enhanced in the interpretations of Foucaultian archaeologies, is not accidental, but essential. It belongs to the very teleological structure of perception, which in the world, always must learn its ways.

And if there are lessons to be learned from the trajectory, they are that we may too easily become caught in the forgetfulness which makes us unaware of our location within the doubled world of today. The doubling, I have suggested, occurs between the literate world and the preliterate world. And there is an enigmatic sense in which questions of primacy cannot occur at this juncture. For if perception was "invented" at the rise of the scientific era, so was writing "invented" at the juncture where humans could begin to discern not only actual, but presumed positions with respect to nonpresent worlds.

In that respect, we contemporaries belong to a doubled world. But from phenomenology we should know that only when variation is possible can there be essential as opposed to merely circumstantial knowledge.

This doubling is, in effect, a multicultural "phenomenological variant." To both "see" in an embodied position, and to "read" in an apparent position, and to be able to easily "hermeneutically" transpose between the two positions is part of what it means to perceive in the now *postmodern* lifeworld. Our perspectives are multiple, refracted, and compound. In that sense phenomenology in its "postphenomenological" moment has recognized something about our very cultural style. This style is simultaneously *both* deconstructive *and* yet structural.

7

Text and the New Metaphysics

I. Introductions: Texts

Today's world of Euro-American philosophy has seemingly immersed itself in the phenomenon of the text. It is as if the virtual totality of things has been subsumed under the paradigm metaphor of the *written,* which in turn relates to a particular activity, reading. Thus we find faddish conference titles such as "Writing the Future," "Writing Woman," or "Writing the Body" abounding. I shall call this contemporary immersion neo-Renaissancism. And my justification for this nomination comes from that marvelous French—rather than Borgesian Chinese—encyclopedia, *The Order of Things* by Michael Foucault.

Foucault rearranges the world according to *epistemes,* sort of atomic Heideggerian epochs of Being, in which discourse is ordered differently in each discontinuous era. In this scheme of things not only are such concepts as "man," "perception," "biology" invented and then disappear, but certain discourses or organizations of knowledge and practice may be fulcral, such as the Renaissance which draws from the Middle Ages, but disappears into the modern era which precedes ours.

And the organization of knowledge-discourse which characterizes this fulcral period is one which has as one of its principles the primacy of *writing.* In short, this is pre-Derrida, Derrida. Hear what Foucault has to say: the sixteenth century is one which is marked by the great metaphor of the book

> which one opens, that one pores over and reads in order to know nature, [and] is merely the reverse and visible side of another transference, and a much deeper one, which forces language to reside in the world, among the plants, the herbs, the stones, and the animals.[1]

This metaphor continues to pervade the beginning of the seventeenth century, Foucault tells us, where "such an interweaving of language and things, in a space common to both, presupposes an absolute privilege on the part of writing."[2] For here is an age also fascinated with its new technologies, including the invention of printing, the arrival from the east of new manuscripts, and a literature not governed by voice, not to mention a *written* music.

> Henceforth, it is the primal nature of language to be written. The sounds made by voices provide no more than a transitory and precarious translation of it. What God introduced into the world was written words. Adam, when he imposed their first names upon the animals, did not more than read those visible and silent marks. . . . Vigenere and Duret both said—and almost in identical terms—that the written had always preceded the spoken, certainly in nature, and perhaps even in the knowledge of men.[3]

Here is the neo-Renaissancism of an "inscription," a "trace" which calls for both a Primal Text and its adumbration into the infinity of interpretation:

> Knowledge therefore consisted in relating one form of language to another form of language; in restoring the great, unbroken plain of words and things; in making everything speak. . . . [Here] the task of commentary can never by definition, be completed. . . . [But] there can be no commentary unless below the language one is reading and deciphering, there runs the sovereignty of an original Text. . . . The language of the sixteenth century [thus] . . . found itself caught, no doubt, between these interacting elements, in the interstice occurring between the primal Text and the infinity of interpretation.[4]

I suggest, in one sense, that the current absorption with the text is thus a return to this fulcral period, although with a difference.

I shall not follow Foucault any further other than to remark that what follows the sixteenth- and seventeenth-century primacy of writing is the modern period which recharacterizes language under a different principle of organization, a principle which in typical semiotic fashion, inverts the previous order. Such an inversion, naturally, means that writing is replaced with speech which becomes primary in the modern era:

> Henceforth, the primary Text is effaced, and with it the entire, inexhaustive foundation of the words whose mute being was inscribed in things; all that remains is representation unfolding in the verbal signs that manifest it, and hence becomes discourse.[5]

Speech inverts writing and commentary is inverted into criticism. Here we see, placed at the seventeenth century, the pre-Derrida/Derrida invention of speech and presence, the echo of which is then attached to Husserl. And given the retrogressive movement of neo-Renaissancism, today we move back from the episteme of speech—and I shall say, *perception* which is also "invented" in the modern era—to the once previous world of the text.

In this playful, but serious, introduction, I have obviously begun to situate myself. And that situation is one which seeks its position at the interstice between writing and the heard, between reading and speech, which is to say, between the text and perception. For I am uncomfortable with the paradigm metaphor of the text, and I am concerned with the almost forgotten role of perception within the lifeworld which I take to be larger than the reduction of it to a textual schema.

II. Ricoeur and the New Hermeneutics

First, I must locate Ricoeur with respect to this scheme of the text and the primacy of writing: what is today "poststructuralism" and deconstruction, yesterday was structuralism. In that recent yesterday, Ricoeur had also entered the fray. The results of those battles are seen in *The Conflict of Interpretations* (French publication 1969, English publication 1974). Under the subtitle of *Hermeneutics and Structuralism,* Ricoeur juxtaposed a series of interpretive processes between an essentially structuralist and a *phenomenological* hermeneutics. In this juxtaposition, structuralism was shown to be a perspective which not only emphasized synchrony over diachrony, but occupied an antipodal position to hermeneutics: "This is why structuralism as philosophy will develop a kind of intellectualism which is fundamentally antireflective, anti-idealist, and antiphenomenological."[6] This stance falls out from "the subordination, not the opposition, of diachrony to synchrony."[7] In short, the problem of temporality in relation to both history and human existence motivated this earlier, but already new hermeneutics.

One of the ingenious points made in this debate was that what could count as privileged examples for such synchronic interpretations was itself only a small range of examples. Thus Ricoeur accused structuralism of having to favor precisely those modes of thought and culture which remained caught in the world of bricolage. "Totemic thought, it seems to me, is precisely the one that has the greatest affinity to structuralism."[8] His counterpart was the ancient biblical, particularly Old Testament tradition which emphasized diachrony, or the series of revelatory *events* which motivated a historically minded people. Thus in the end, "bricolage works with debris; in bricolage the structure saves the event; the debris plays the role of a preconstraint, of a message already transmitted."[9] But within a basically temporalistic tradition such as the biblical one, "the reuse of biblical symbols in our cultural domain rests, on the contrary, on a semantic richness, on a surplus of what is signified, which opens towards new interpretations."[10] Here temporality linked to history is already a primary factor in the new hermeneutics.

Today, Ricoeur continues the debate within the context of today's rampant *textuality*. But his approach is again instructive. It is reflected in the title of the recent *Time and Narrative* volumes. It is not accidental that the term "narrative" is already suggestive of a more actional, dynamic process than the term "text." For in this nuanced difference there already lies the role of a revised phenomenological hermeneutic which places itself within the current debate, and in such a way that its forgotten dimensions are again taken up.

What Ricoeur restores to the debate is the emphasis and base of all writing within the context of a human-world relationship. But an interrelation between the human and the world across *temporality* and through *reference*. Thus when he takes up the question of narrative, it is in the context of prefiguration through *configuration*, which is to say that there is a structure of narrative, through a *refiguration* which is a temporal movement which redoes in an actional way any previous configuration. Here we have the maintenance of the essentially gestalt model of a phenomenological analysis, but it occurs also in terms of that which points beyond itself in terms of *reference*. At the end of volume 2, Ricoeur provides the following summary:

> The problem of the refiguration of time by narrative will, therefore, be brought to its conclusion only when we shall be in a position to make the referential intentions of the historical narrative and the fictional narrative interweave.[11]

Here, just as in the earlier debates with structuralism, Ricoeur has juxtaposed himself against the implicit internalism of poststructuralism. There remains, for Ricoeur, *referential intention.* But this referential intention is not anything like a sentential one, it is rather a gestaltist one which becomes clear only through the multiple dimensions of metaphor (*La metaphor vive*) and through the interrelation of the existent human and a world:

> Our analysis of . . . time will at least have marked a decisive turning point . . . by providing something like a *world of the text* for us to think about, while awaiting its complement, the *lifeworld of the reader,* without which the signification of the literary work is incomplete.[12]

Narrative is both actional and structural. As a plot it provides structure, meaning, but in the movements from configuration through refiguration in both historical and fictional narrative, there is revealed the constitution of textual worlds in correlation with human lifeworlds.

Because the purpose of this paper is not simply Ricoeur exposition, I shall not undertake a long excursus, but shall rather suggest several of the threads of continuity and modification which go to make up this new hermeneutic:

1. Echoing the relation between experience and language found in the *Symbolism of Evil* in which those primitives of the experience of suffering and evil find expression in an already doubled expression (the analogs of stain, sin, and guilt), narrative, which is *muthos,* also interrelates experience and now narrative expression:

> My basic hypothesis [is] that between the activity of narrating a story and the temporal character of human experience there exists a correlation that is not merely accidental but that presents a transcultural form of necessity. . . . time becomes human to the extent that it is articulated through a narrative mode, and narrative attains its full meaning when it becomes a condition of temporal existence.[13]

2. The task of eliciting this connection is necessarily hermeneutic rather than structural. In short, here is again the contrast between the limits of a semiotics in contrast to a lifeworld semantics:

> It is the task of hermeneutics, in return, to reconstruct the set of operations by which a work lifts itself above the opaque depths of living, acting, and suffering, to be given by an author to readers who receive it and thereby change their acting. For a semiotic theory, the only operative concept is that of the literary text. Hermeneutics, however, is concerned with reconstructing the entire arc of operations by which practical experience provides itself with works, authors and readers.[14]

3. Often now citing the notion of the Husserlian lifeworld, and its use of a genetic, i.e., historical-ontological derivation, the movement of time and narrative is from "the destiny of a prefigured time that becomes a refigured time through the mediation of a configured time."[15] At the level of prefiguration is the realm of concrete human action, situated within whatever historical or fictive context is referred to. Here is the ground level of agents, with actions, having motives. "There is no structural analysis that does not borrow from an explicit or an implicit phenomenology of 'doing something.' "[16]

4. There is also a modification of the usual noematic-noetic analysis of phenomenology in the Ricoeurean approach to texts and narrative, in that any theory of writing must relate correlatively to a theory of reading. It is here that the problem of reference in its narrative context arises:

> An aesthetic of reception cannot take up the problem of communication without also taking up that of reference. What is communicated, in the final analysis, is, beyond the sense of the work, the world it projects and that constitutes its horizon.[17]

But the "world of the text" is not, and cannot be equivalent to *the* world. Here is where Ricoeur distinctively breaks with the current penchant to reduce world to text:

> Reference and horizon are correlative as are figure and ground. All experience both possesses a contour that circumscribes it and distinguishes it, and arises against a horizon of potentialities that constitutes at once an internal and external horizon for experience.[18]

But, in this interplay Ricoeur also argues that "language does not constitute a world for itself. It is not even a world. . . . Language is for itself the order of the Same. The world is its Other."[19]

5. But the correlation a priori of human and world is one in which "all reference is co-reference."[20]

> What a reader receives is not just the sense of the word, but, through its sense, its reference, that is, the experience it brings to language, and, in the last analysis, the world and the temporality it unfolds in the face of this experience.[21]

6. Finally, because this form of referentiality is not merely literal or descriptive, particularly in fictional and historical narrative, it refers through *complexity*. One might say that the "primitive" of narrative is precisely this complexity, the complexity of what Ricoeur previously studied as metaphor. "I tried to demonstrate in *The Rule of Metaphor* that language's capacity for reference was not exhausted by descriptive discourse and that poetic works referred to the world in their own specific way, that of metaphorical reference."[22]

What we arrive at through this hermeneutic process is the already mentioned correlation of the world of the text with the lifeworld of the reader. Thus, in a fictional text, "what is interpreted in a text is the proposing of a world that I might inhabit and into which I might project my own most powers. . . . Poetry (or fiction) through its *muthos* [narrative], redescribes the world."[23]

Here, then, is a brief sketch of a new hermeneutics addressed to the problem of time and narrative, within which the role of the reader in relation to the text is a relation between the lifeworld of the reader and the world of the text. Its framework is phenomenological, now modeled after, but dynamically creviced from, the Husserlian concept of an actional lifeworld situated between the historical and the imaginative within the movements of refiguration, configuration, and prefiguration.

I probably need not say that I find this approach refreshing in the contemporary world of deconstruction and hermeneutics. Ricoeur seems somehow to always address that which is most contemporary, and yet at the same time retain his sense of balance. For that I am grateful, and from that I have often learned.

But if Ricoeur has been able to reformulate a phenomenological hermeneutic to address the textuality of the contemporary scene, there are also aspects of a relation between phenomenology and the phenomena of writing and reading which have *not* yet been addressed. And it is here that I shall take up my own position within this discourse.

III. Perception, Reading, and Writing

What has been left out, to my mind, is precisely what could be called a *phenomenology of reading and writing.* At best it has been implicit and operational behind the scenes. What I propose to do is to open precisely the field of this phenomenology of reading and writing, and to do it by way of reintroducing the role of *perception,* or better, perceptual action, into the arena.

One could, of course, recall both the Husserl of the lifeworld period here, as well as Merleau-Ponty's primacy of perception. For Husserl, in his analysis of the Galilean world, argued that it was a break with the basic lifeworld which, at bottom, must refer back to the ordinary perceptions and actions of humans, an echo of which we have found in Ricoeur's position. Similarly, although somewhat more subtly with respect to the interpretation of perception, one finds the analyses of Merleau-Ponty regarding motility, perceptual gestalts, etc., particularly in the worlds of painting and the arts.

I shall begin by taking note of that most extreme deconstructionist, Jacques Derrida, by suggesting that within the techniques of interpretation and deconstruction he so frequently employs regarding seemingly odd or distorted readings, there lurks an *implicit* phenomenology of perception. I noted in an earlier article, "Phenomenology and Deconstructive Strategy," that such an implicit phenomenology lies within Derrida's treatment of texts:

> Take a text: If one views a text (perceptually) it usually appears first as writing which is centered on the page, surrounded by margins, but the focal center is clearly the bulk of what is written. Then, if one reads the text, what usually emerges as focal is what the text is about, however complex that may be, as indeed any text usually is. What does Derrida do with a text? Posed in the way I have indicated, he immediately decenters what seems to be focal and immediate. His focus is radically shifted to titles, signatures, margins, borders, divisions, etc. In short, he draws our attention to features which are there, but usually taken at most as background, secondary or unimportant features.

> In a sense this is a highly "phenomenological" technique. For example, in an analysis of perception, phenomenologists like to point out that while what stands out (figures) are usually most obvious because they are the referenda of our usual perceptions, all figures take their position upon a background which is equally present and which constitutes the field of perceivability. In short, this move "decenters" focal perception so as to attend to taken-for-granted but important fringe features. Similarly, to point out that all perceptions include not only manifest surfaces, but latent "backsides," is to "decenter" at least the usual interpretations of perception. I am suggesting that this device—perhaps taken to Nietzschean excess—is a familiar ploy of Derrida. Indeed, one can see, once the operation is known, how to follow along with such deconstructions. (Is there a Derrida text which addresses itself to the empty background of the page? If not, there ought to be.)
>
> Once the focus is decentered, however, a second more radical step is taken. In the Derridean tactic this seems to be a kind of playfulness which then wants to read the fringe back into the center as a kind of shadow presence. This tactic appeared early in the assertion of the primacy of writing in *Of Grammatology* and *Writing and Difference.* If writing "precedes" speech (decentering the subject), its "evidence" is the trace which must be found but since the trace is not obvious the finding calls for a kind of play. Similarly in "late" Derrida, the margins, signatures and borders get played back into the focal text. It is in the brilliance of this playing that Derrida becomes genius-magician with his supporters noting only the genius and his enemies only the magician.[24]

This, however, is to *indirectly* broach the problem of a phenomenology of reading and writing. Rather, now I would like to develop a different example-set by looking at an interesting historical illustration of the role of perception and bodily position in relation to the phenomena of reading and writing.

I have deliberately chosen an example in which there is the establishment of a *reading perspective* which is then implicated in a large praxical perspective. The example relates to the reading of charts, particularly navigational charts such as are used by seamen. I shall then associate this "microperception," i.e., actual bodily position and perceptual perspective, with what I shall call a "macroperception" or the interpretive praxis which permeates a larger set of cultural perspectives.

As early as the fourteenth century there began to appear a new set of maps or charts in Europe called *portulans*. These were basically what today would be called pilot charts, or maps which display in a visually isomorphic way the coastal and land configurations along some sailing route.

These charts were different from the also then popular Mappae Mundi which displayed a different picture of the world. The "world maps" of the medieval imagination displayed an essentially mythical world which had the shape of a round earth, crossed by two rivers in the shape of a "T." These maps located Jerusalem as the center of the earth, were land-oriented, and usually filled with mythological characters and locations—they obviously could not be used for any voyaging or navigation! The portulans, in contrast, were among the first "empirical" maps and carefully traced routes along coasts and between land masses. They were sea-oriented and routes were also laid out in terms of an imposed "grid"—only in this early case, the grid was not yet that of today's longitude and latitude grid, but of the dominant *winds*. Thus, upon the face of the earth, there were guidelines for sailing, the grid of the winds (all of which were named in those days).

Note, now, two other features concerning these maps, features which in reading seem almost too obvious: (1) first, the map was an abstract, but essentially isomorphic map regarding the shape of the coastline, land masses, etc.; but (2) the isomorphism was displayed to a reading perspective, to a position *above* the earth itself. These features we take as so obvious, so taken-for-granted, that it never occurs to us that such a representation has a certain oddness. That oddness is one which is experienced by the actual navigator—for he (or she) does not actually see the landmass or coastline in this fashion. Rather, the bodily position of the actual navigator is one of being located upon the deck or vessel. Thus, from *reading* the chart to finding one's position from a *lateral* perspective on board the boat, there must be added a particular hermeneutic act, the act of transferring a perspective and position from overhead to lateral.

In short, with the portulans, we have the constitution of an "empirical" view of the world, but constituted from the noetic position of the "god's-eye" perspective, from above.

Now this is, in fact, a normal situation for *reading*. And this obtains with respect to reading other "texts" than those of charts. That is, normally, we sit, with book in front of and usually below our eyes, or, as was quite normal in the Middle Ages, standing, reading from above. In that respect there was already a sedimented practice regarding the reader/text position with relation to a bodily perceptual stance.

The isomorphism of representing an actual coastline introduces a reading "realism" into the process which has stronger implications for both a privileged perspective and for the associated hermeneutic acts of interpretation needed to fill out the navigational praxis.

To make this feature stand out with even more distinctness, let us do a few phenomenological variations on map-reading, first by varying the noematic or object correlates: (1) maps need not be isomorphic nor overhead. One could, as is also done in directing someone to one's house, have given a set of written directions: "You take Hollins Road for about a mile, until you hit the Griswold Pub; then right until you get to Wistful Cottage, my place." Here the positionality of the reading is not so important since isomorphism is not part of the process, but there is the hermeneutics of reading which must recognize from a written description of the features of a perceived landscape; or (2) one could have as part of the navigational map (and as, fortunately, is common in British circles) a representation of a lateral perspective either separately or added to the regular chart. In this case the position of the navigator on board corresponds to what is seen in the chart. "The two towers, when lined up, serve as range markers for the entrance to Swipshold Harbour." Here the all too frequent ambiguity suffered in trying to interpret where one is when approaching an unfamiliar landfall, is eased. In the overhead perspective, I must imagine that towers belong to what chart representations, and what they must look like to someone in a lateral position.

This simple example, so familiar, should be expanded. Take what is equally familiar to the inhabitant of a literate culture, the representation of any number of landscapes from elevated or bird's-eye perspective: I have a large book on the history of gardening and from very early times one finds representations which are from elevated or overhead perspectives. These seem to be cross-cultural insofar as they occur in *literate* cultures. Thus, medieval gardens, Renaissance gardens, Chinese, Japanese, Indian and Persian gardens are likely to be displayed from overhead or elevated perspective, perspectives which are rarely actually occupied by the gardener.

The same, however, is not the case with respect to many *nonliterate* cultures! Instead, representation will frequently be lateral, often not even in perspectival form at all. Thus, in certain South American bark paintings, one finds simply a field of scattered figures (humans, animals, canoes, etc.) displayed on the painting, but each to be seen as from the side, laterally. Or, if organized, the laterality may take the shape of some series of laterally displayed figures, as in some Australian aboriginal bark paintings. In short, the establishment of a coherent or "empirical" display, and the favored god's- or bird's-eye implied position seems to be lacking.

But is it lacking? Or is this the trace of a different seeing of the world?—I am reluctant here to call it a "reading" of the world, since reading in a nonliterate culture will be somewhat metaphorical, or perhaps projected from ours.

I return briefly to navigation, this time with an example from a nonliterate culture, from a culture without maps. I refer to the equally or even more ancient practices of South Pacific navigation, the principles of which have only recently been reconstructed by Westerners. Note that historically, the South Pacific peoples successfully navigated and populated virtually the entirety of the Pacific in times equal to or more ancient than our Middle Ages. Thus their methods were clearly pragmatically successful.

But they did this without maps, and, as I shall show, without utilizing our praxis of overhead perspective. How? I shall call their system a *perceptual hermeneutics.* We again might be tempted to say that they "read" nature as their "text." But again this is to take our own presuppositions and cross-culturally impose the habits of a reading or literate culture into these perceptions. Let me schematically indicate several features of this perceptual hermeneutics to illustrate what is involved.

1. Once a pathway through the sea was known, there were methods of retracing those pathways. Starpaths were very well known to these navigators. From childhood the boys would canoe with their elders and the stars—which were interpreted as pillars of the skyhouse—were named and known through canoeing songs. Then, once identification was learned, the various formulae, not unlike the instructions I cited earlier in telling one how to reach my cottage, were given in other verses of these songs. "If you wish to reach Tahiti, you follow the Great Eastern Bird until the ghost island, Mapu, reaches you, then you turn and follow the Fish, etc." These memorized formulas have persisted to the present and David Lewis, the anthropologist-sailor who has studied these methods the most, was able to get Hawaiians to retrace a voyage from Hawaii to Tahiti which had not been made for three hundred years—without instruments or charts!

2. Another feature is the need to be able to steer a straight course, often in reduced visibility or even in storms. These navigators learned to follow direction by becoming kinesthetically sensitive to the dominant swell patterns of the Pacific. The trade winds create long parallel swells which persist through the sailing season, and which can be detected even

under or in the midst of confused local seas. That is, they can, if the navigator sits down into the belly of the catamaran and feels them with his testicles!—so say the Polynesians.

Moreover, these swell patterns also refract when approaching an island, and the refraction is one of the clues to finding the island. Thus, like listening to the bass rhythm of a musical piece, the detection of swell patterns provides perceptual clues to direction.

3. And how does one spot an island, particularly if on a voyage of discovery? The answer is again perceptual: islands with mountains and jungles (a) cast a greenish glow upon the sky for many miles away, (b) closer, the clouds which collect over an island landmass do not move—even though higher clouds are moving—and may be spotted from great distances. (c) Birds at dusk return to their roosts, with some species beginning the return from fifty miles out, others from thirty miles out, etc. The navigator knows all the birds and notes their dusktime direction.

And so it goes, a subtle, perceptual hermeneutics, utilizing neither map nor overhead perspective. Indeed, range marking, even the abstraction of "ghost islands" are all described from lateral perspective. But there is one more, and especially striking, feature which helps sediment this system of navigation as an *embodied position* system. On board, as with many seagoing terms and languages, the language of spatial orientation and position changes. It becomes one in which the fixed variable needed for this relativistic system becomes one's actual bodily position. Thus, one does not speak of "going to" Tahiti, rather, one speaks of Tahiti "approaching us" or "coming to us." For the boat also does not "go through the sea," rather the "sea passes by us." Here is a primitive "Einsteinian" world in which whatever is measured must take into account the position of the actual observer.

If such a perceptual system might seem strange to us, it is when we begin to compare the two systems that even more interesting factors emerge. For example, if I am able to sympathetically enter the perceptual world of the Pacific Navigator (did I ever leave it? since I do live my body and am positioned with respect to the world, even if I overlook the dominance of laterality in ordinary action), then when I ask "from what position could I have established that my canoe is going to Tahiti?" the answer would be, "from the fixed 'god's-eye' view which is above me, but which I do not actually occupy." For at sea, when the horizons are in motion, when the bow wave passes the bow, it is perceptually relative to say that the sea is passing by my craft, or I am passing through the sea. But both statements reflexively imply a position from which the statement is "true." Thus in this now cross-cultural example of a variation, I begin to discern the primacy, in my own culture, of an imagined or quasiperspective as the privileged one.

Yet it is a perspective which today we take for granted, which is familiar and sedimented. Nor do we note its strangeness or oddity. Could it be that it in any way implies a relation between our bodily microperceptions which, as Merleau-Ponty so brilliantly demonstrated, we forget in distancing the lifeworld and the scientific world, and the macroperception of a privileged transcendent view? I would like to suggest that what I am referring to is the sediment of a long practice of *reading*.

Or, in another way of putting it, can there be in our practice of being literate, not only some relation between the world of the text and the lifeworld of the reader, but a deeper implication even for the bodily, perceptual position in which we find ourselves when we read a text, any text? Is the god's-eye view a projected and privileged reading position?

IV. Conclusion

There must be some connection between bodily position and perception; its extension and extrapolation into the macroperceptions of a culture has been suspected for some time in phenomenology. When Merleau-Ponty claimed that culture is perceived, I am suggesting that there is something stronger than metaphor involved. But what of the interconnection between reading position and this macroperception—is there evidence for this reading-perceiving which constitutes our praxis?

My last example will make this suggestion stronger. Ancient Egyptian bas-reliefs often had odd perspectives. They were perspectives which in the same panel combined overhead and lateral perspectives, perspectives which with respect to their implied observer's position could only appear to be "inconsistent." Thus, in the depiction of a garden, there could be a laterally perceived series of vines being tended by the workers, but in the same scene, the water pool would be depicted from a directly overhead position. Here was a "doubled" perspective which, when thought about seems odd, yet which must have seemed quite conventional and familiar to the ancient Egyptian.

But if one now switched to the reading which was practiced by the Egyptians, suddenly one discerns the same phenomenon. Hieroglyphics were a mixed or "doubled" writing, part pictograph, part phonetic. Thus, while it might be an ibis or a bull which was depicted, its significance could be either what it represented, or it could be a phoneme sounded within a word or sentence: a "doubled" text with respect to reading. Thus, in both representation and in the written text, Egyptian seeing-reading was doubled, constituting a familiar praxis within which life was ordered.

This was a lifeworld in which the polymorphy of perception took a particular shape, even when unnoticed and even in the midst of one of the early cultures which could be called literate. I suggest that the same praxis, although shaped differently, orders our lifeworld, and thus while it makes a good deal of sense to utilize the reading metaphor widely, it does not make sense to forget its involvement in the unspoken and unwritten domain of perception and bodily action within a material world.

8

Deromanticizing Heidegger

A century after his birth, two very contrary statements can be made concerning Martin Heidegger: (1) In a significant sense, he is surely one of the most important founders of the philosophy of technology. His insights into the structures and functions of technology remain deep and suggestive. (2) But we all also know that he joined the National Socialist party and remained with it through the war. His associations with the movement, seen today as one of the most destructive applications of modern technology, are equally deeply disturbing.

My question is: is there something at the very heart of Heidegger's thought which makes both of these contraries possible? And I begin my reflection with two vivid images, both related to that ancient Greek ancestry to which Heidegger turned again and again as a source of thinking, consonant with self-proclaimed origins for Euro-American civilization.

The first image is Heidegger's, that of the famous Greek temple image in "The Origin of the Work of Art." Heidegger's temple is taken as a paradigm of artful *technē,* both "thingly" and signifying.

> Standing there, the [temple] rests on the rocky ground. This resting of the work draws up out of the rock the mystery of the rock's clumsy yet spontaneous support. Standing there, the building holds its ground against the storm raging above it and so first makes the storm itself manifest in its violence. The luster and gleam of the stone, though itself apparently glowing only by

> the grace of the sun, yet first brings to light the light of the day, the breadth of the sky, the darkness of the night. The temple's firm towering makes visible the invisible space of the air. The steadfastness of the work contrasts with the surge of the surf, and its own repose brings out the raging of the sea. Tree and grass, eagle and bull, snake and cricket first enter into their distinctive shapes and thus come to appear as what they are.[1]

This "Wagnerian," this "Nietzschean" deployment of signification in the focal point of the Greek temple against the ground of its earth is typical Heidegger. It is heavy; it is *romantic*; it is what gathers the mortals, gods, earth, and sky. "The temple-work, standing there, opens up a world and at the same time sets this world back again on earth, which itself only thus emerges as native ground. . . . The temple in its standing there, first gives to things their look and to men their outlook on themselves."[2]

Heidegger's imagery is striking, captivating, and, above all, weighty. Now contrast what could be another look at the same image, done this time by a historian, J. Donald Hughes, in an obscure book, *Ecology in Ancient Civilizations.*

> Those who look at the Parthenon, that incomparable symbol of the achievements of an ancient civilization, often do not see its wider setting. Behind the Acropolis, the bare dry mountains of Attica show their rocky bones against the blue Mediterranean sky, and the ruin of the finest temple built by the ancient Greeks is surrounded by the far vaster ruins of an environment which they desolated at the same time.[3]

Here, the same "thing," the Greek temple, reveals a very different "world" from that of Heidegger. Hughes goes on to point out:

> In the centuries before the Golden Age of Athens, those same mountains were covered by forests and watered by springs and streams. The philosopher Plato saw evidence of the changes that had occurred not long before; there were buildings in Athens with beams fashioned from trees that had grown on hillsides which by his day were eroded and covered only with herbs, and he visited shrines once dedicated to the guardian spirits of flowing springs which had since dried up.[4]

What accounts for the dramatic difference in what is seen in the image of the Greek temple? Or, phrased even more starkly, is there, in the Heideggerian way of seeing, a deeper and even necessary way of *concealing* which allows only a romanticized perspective?

I opened this reflection with a well-known example from one of Heidegger's analyses of art objects. But his analyses of technological objects follow a close and similar pattern. Indeed, one could easily conclude that an art object is, for Heidegger, the primary example of a "good" technology. Both art objects and technological objects—equipment—are "thingly," "produced," have ways of "revealing" a world, and belong in some way to the process called *technē*, which Heidegger defines in the following passage:

> There was a time when it was not technology alone that bore the name *technē*. Once that revealing which brings forth truth into the splendor of radiant appearance was also called *technē*. Once there was time when the bringing-forth of the true into the beautiful was called *technē*. The *poiesis* of the fine arts was also called *technē*.[5]

One cannot but detect, again, the heavy romantic overtones of this nostalgic merging of art and technology. Nor should we ignore from the outset that Heidegger's primary suggestion of a solution to the dilemmas of the Age of Technology often revolve around a kind of saving power found in art. But if this is where the revealing power that could save us from the reductions of modern technology is, it is because art and technology are closely related in precisely the thingly, produced, but revealing roles which both art objects and equipment or technological objects contain when they are seen as focal elements against a context or field which is "lighted up" as a "world."

Yet, in the Heideggerian corpus, there is often a great difference of evaluation and connotation between art objects and technological objects. On the surface, it might appear that the two most frequently patterned such differences relate to a certain suspicion concerning *modern* technology versus traditional technologies, and the older, smaller and simpler technologies versus the newer, larger and more complex technologies.

There is much in the Heideggerian choice of "good" and "bad" connotations which commentators have noticed. Heidegger "likes" the tools of the workshop, the peasant shoes of the Van Gogh painting, the watermill on the stream, the windmill, and the old stone bridge with its arches. He does not like hydroelectric dams on the River Rhine, the atomic bomb, even the modern steel bridge which routes traffic to the same city square as the old stone bridge. Such a pattern would seem to evidence a simple and old fashioned romanticism of a nostalgic sort—nor would I deny that such a strain my be found in Heidegger. But the issue is more complex than that.

There are, for example, inconsistencies: it is not always the small and relatively simple which escapes the Heideggerian disapprobation. Clearly, by today's standards the typewriter would be an example of a good, simply mechanical writing device, not much different in kind or principle than the Bavarian clocks beloved by the folk of Todtnauberg. Yet the typewriter receives a particularly scornful disapprobationary note:

> It is not by chance that modern man writes "with" the typewriter and "dictates"—the same word as "to invent creatively"—"into" the machine. This "history" of the kinds of writing is at the same time one of the major reasons for the increasing destruction of the word. The word no longer passes through the hand as it writes and acts authentically but through the mechanized pressure of the hand. The typewriter snatches script from the essential realm of the hand—and this means the hand is removed from the essential realm of the word.[6]

I virtually feel the scorn which would have been poured upon my composition of this paper with a word processor! But the point is that here is a relatively simple, mechanical device which does not escape the romantic thesis. I shall return to this example.

There are also much deeper inconsistencies in this pattern of choices of "good" and "bad" technologies. In the "Question Concerning Technology," a deep danger of modern technology is laid to the way the world is revealed in the ensemble of modern technology as "standing reserve," the extant translation of *Bestand.* (I translate this term as "resource well.") It is the whole of nature which is revealed as a resource well in modern technology, as illustrated through the disliked hydroelectric plant on the Rhine:

> The hydroelectric plant is set into the current of the Rhine. It sets the Rhine to supplying its hydraulic pressure, which then sets the turbines turning. This turning sets those machines in motion whose thrust sets going the electric current for which the long-distance power station and its network of cables are set up to dispatch electricity. In the context of the

> interlocking processes pertaining to the orderly disposition of electrical energy, even the Rhine itself appears to be something at our command.[7]

That is, nature, including the Rhine, is revealed as resource well, standing reserve, for man's use. And while this production of energy does contrast with the old windmill—which can turn only when the wind blows and which thus seemingly lets the wind "be"—it does not, in principle, differ from the smaller dam on the stream which allows the water-wheel in turn to grind the peasant's wheat. To allow this example as a "good" technology does not, to my mind, prevent seeing nature as resource well except in its lack of a larger interconnectedness with the electrical grid.

To this point, it should be clear that the romantic thesis, as I shall call it, pervades Heidegger's choices of "good" and "bad" technologies. But in what does it consist?

The first element, I claim, is a preference for what I call *embodiment relations.* Heidegger prefers, likes, those technologies which express straightforward bodily, perceptual relations with the environment.

This is part of what underlies his dislike of the typewriter. Expressivity, the connection with "word," is primitively found in a handy gesture:

> Human beings "act" through the hand; for the hand is, like the word, a distinguishing characteristic of humans. Only a being, such as the human, that "has" the word (*mythos, logos*) can and must "have hands." . . . The hand becomes present as hand only where there is disclosure and concealment. . . . The hand has only emerged from and with the word.[8]

As we saw with the typewriter, for Heidegger somehow there is less "hand" in writing with a typewriter than presumably that which is "handwritten" with a pen.

This same preference for simple, embodiment relations is exemplified in his very earliest analysis of human-technology relations, the famous "hammer example" in *Being and Time.* That example is simultaneously one of the most pointed in showing not only Heidegger's radical insights into technology but also a certain blindness and prejudice concerning technologies which do not express embodiment relations.

I shall not here go into great detail into this often analyzed example but shall note only a few salient features which are relevant to the way in which the romantic thesis also conceals important aspects of technology.

Positively, Heidegger shows in the hammer example that technologies in use are not objects as such; they "withdraw" in use and become partially transparent means by which humans relate to an environment. Here is a good critique of any simplistic and objectivisitic account of technologies as simple objects. Rather, technologies are contextual, or field involved; the hammer "is" what it is in reference to the context of nails, project, etc. It belongs to a reference system which always includes more than a mere hammer. Thus, while the hammer is always "thingly," it is never a *mere* thing and is, in use, transformed into a world-related and world-revealing way in which humans are involved with their environments. All of this—and more—is the source of the Heideggerian suggestivity for philosophy of technology.

But there is also a negative side to the analysis. In *Being and Time,* the context is "lit up" through technological *breakdown.* It is when the hammer is broken or missing that its involvements are shown. The fullness of the project—and the objectness of the hammer—get shown when it is *not* functioning. I claim that here lies an early clue to a certain negativity which pervades the Heideggerian corpus and which blinds the analysis both to a possible appreciation of human-technology relations other than embodiment ones and to the features which, in fact, unite modern technologies to traditional ones. In *Being and Time* it is hard to conceive of a positive relation *to* a piece of equipment, a technology, other than as that *through which* Dasein experiences its environment either in embodiment or with transparent referentiality. (The old turn-signal example from the German automobiles which had flip-up arrows manually operated is what I call a *hermeneutic relation,* since actions are "read" *through* the technology. These are recognized in *Being and Time* but are on a close continuum with embodiment relations and are directly expressible in the technologically mediated action of the driver.)

In short, to relate *to* a technology in a positive way and in a situation in which the artifact takes on what I call an *alterity* relation seems to me inconceivable in the Heideggerian scheme. And although I cannot long belabor this here, an example from technology-as-toy may illustrate what I have in mind.

The child's top is just such a technology-as-toy which may become an *alterity relation.* Set in motion, the technology itself becomes an object of fascination. It has a quasilife of its own, even apparent self-movement which is unpredictable. It becomes a quasi-other to which the child can happily relate. Such playful technological moments do not seem to belong to the heavy romanticism of the Heideggerian context. But just for that reason one could also miss the kind of appreciation and fascination which characterizes much of the experience of modern technologies.

The preference for embodiment relations over other human-technology relations is what could be called a nostalgic element in the romantic thesis. It is hardly unique to Heidegger. It is also to be found in Karl Marx. Insofar as alienation theory is bound to any nostalgic element relating to the handwork of the worker prior to machine tools in a factory context, there may be found the same taste preference in that older mode of analysis.

A second element of the romantic thesis is one which has made Heidegger so appealing to those close to the environmental movement, particularly those now associated with "deep ecology."

Here, too, a certain closeness between "good" technologies and the art object emerges more clearly.

A "good" technology is one which "gathers" a world in a certain way and "lets be" the nature and community which is so gathered. The old stone bridge in the essay "Building, Dwelling, Thinking," is an excellent example of this subthesis. Like the temple, the bridge reveals a certain world:

> The bridge swings over the stream "with ease and power." It does not just connect banks that are already there. The banks emerge as banks only as the bridge crosses the stream. The bridge designedly causes them to lie across from each other. . . . It brings stream and bank and land into each other's neighborhood. The bridge *gathers* the earth as landscape around the stream. [And in a direct echo of the temple, the] . . . waters may wander on quiet and gay, the sky's floods from storm or thaw may shoot past the piers in torrential waves—the bridge is ready for the sky's weather and its fickle nature. . . . The bridge lets the stream run its course and at the same time grants their way to mortals so that they may come and go from shore to shore. . . . The bridge *gathers* to itself in *its own way* earth and sky, divinities and mortals.[9]

Here art object and use object merge positively. The bridge is clearly a "good" technology when it gathers the Heideggerian fourfold in its focal/field relation. Yet, not all bridges do this, according to Heidegger:

> The highway bridge is tied into the network of long-distance traffic, paced as calculated for maximum yield. [This stands in contrast to the] . . . old stone bridge's humble brook crossing [which] gives to the harvest wagon its passage from the fields into the village and carries the lumber cart from the field path to the road.[10]

The steel highway bridge is to the old stone bridge precisely what the typewriter is to the pen. Yet even in the midst of this clearly romanticized difference, Heidegger must admit that both bridges have something in common: "Always and ever differently the bridge escorts the lingering and hastening ways of men to and fro, so that they may get to other banks and in the end, as mortals, to the other side."[11] This means more than that a bridge functionally connects the two sides of the river. It means that a bridge, ancient stone or modern steel, *gathers* in its way. Authentically or "good" sounds in the term "lingering," whereas unauthentic or "bad" in the "hastening," which belongs to the highway network. So, gods and mortals, sky and earth are gathered—although differently—by both bridges: "The bridge *gathers,* as a passage that crosses, before the divinities—whether we explicitly think of, and visibly *give thanks for,* their presence, as in the figure of the saint of the bridge, or whether that divine presence is obstructed or even pushed wholly aside."[12]

Gathering, the deep signifying surplus of meaning in the figure/ground of bridge and landscape, thus belongs to any such artifact. The difference lies in Heidegger's late variant upon how this may happen authentically or unauthentically. And while the romantic thesis clearly belongs intimately to this distinction, it does so in a complicated way.

The authentic mode of gathering is one which states, "To preserve the fourfold, to save the earth, to receive the sky, to await the divinities, and to escort mortals—this fourfold preserving is the simple nature, the presencing, or dwelling."[13] And it is at least implicitly clear that the conservative view of Heidegger's notion of Germanic life is what best fulfills this preservation. That, too, is part of the romantic thesis. The authentic mode of gathering contrasts with its unauthentic counterpart, the world as revealed through the "gathering"—which unauthentically becomes *enframing* in the worldview of modern technology.

I am quite aware that Heidegger does not simply outright condemn modern technology—its essence, enframing, is simultaneously a revealing of the world and an openness:

> When we consider the essence of technology we experience enframing as a destining of revealing. In this way we are already sojourning within the open space of destining . . . [and] when we once open ourselves expressly to the *essence* of technology we find ourselves unexpectedly taken into a freeing claim.[14]

But there is danger here of the unauthentic closure and reduction which arises from this form of gathering: "The coming to presence of technology threatens revealing, threatens it with the possibility that all revealing will be consumed in ordering and that everything will present itself only in the unconcealedness of standing-reserve."[15] In short, all of nature, including human being, will be seen as reduced to a vast resource well—but the question then is: for whom, or for what end?

So, the stone bridge or the steel one, the temple or—and here I shall introduce a counterexample bound to be outrageous to Heideggerians—the nuclear plant gather the fourfold, albeit in different ways. Here, then, is my post-Heideggerian example:

> Seen while sailing in Long Island Sound, on the horizon stands the stark super-silo, light green topped, of the Shoreham nuclear plant. Standing there, it brings to presence the very contrast between the seemingly featureless sandhill earth with the sky. It stands at and defines the contrast, too, between the sea and the shore, which without its focal presence would also be featureless lines along the horizon.

I could, of course, go on in this crypto-Heideggerian mode, but the point is made. But if the nuclear plant substitution for the temple is somehow outrageous, what makes it so? I contend that the difference is not simply the difference between the nostalgic romanticism of the Greek temple and the urgent and fearful presence of the nuclear plant. Rather, it lies in what is left out, concealed, or unsaid in the Heideggerian account.

What is left out of both the Heideggerian account of the temple—and of the crypto-Heidegerrian paraphrase concerning the nuclear plant—is what Langdon Winner has called the "politics of the artifact." For us, that dimension of the thingly is more vividly present in the nuclear plant than in the lost civilization of the Greeks only because it is nearer to us.

We know that the Shoreham nuclear plant was sold this year to New York State for one dollar (its cost had been five billion dollars). We know that the reasons had to do with the recognition that Long Island, the most populous nonurban area in the United States, could not be evacuated in a Three Mile Island- or Chernobyl-type accident, and with the political opposition of over 75 percent of the populace to its going on line.

Yet, ironically, precisely because the nuclear plant *revealed* its artifactual politics in a somewhat Heideggerian fashion, this first stoppage of a plant before opening could occur. It was because the Shoreham plant revealed—negatively—its form of gathering that it could be closed. So, as at the beginning of this reflection where I juxtaposed J. Donald Hughes's reading of the Greek temple with Heidegger's, the same can be done here: In its gathering, the nuclear plant makes the fishy life of the Sound to appear—as drawn to the warmer exhaust waters of the plant, but to be placed in danger of a leak, as in the case of the Irish Sea, which today is the world's most radioactive sea because of the irreparable leak at Folkestone. It channels the community into its pathways—now recognized to be more cloggable than the Long Island Expressway at high traffic time. It reveals the hastening which would be needed to evacuate its wastes (by sea, or by land?), etc. And I add as a postscript to this Hughesian variant upon Heidegger that I could not borrow the even more poignant image of Langdon Winner in his description of the Diablo Canyon plant, beyond which he sighted a gray whale in the Pacific. This is because Long Island Sound has ceased to entertain whales or porpoises for the entirety of the twentieth century and, as of 1987, has its first forty miles of bottom so hypoxic (absent of oxygen) because of excessive phosphate and nitrogen runoff from modern sewerage that even the lobsters are dead in that area.

By adding this politics of our artifacts to the analysis—absent from the Heideggerian account, concealed and unsaid—the account becomes even more powerful. What needs to be noted, however, is that the romantic thesis in its unsaying concealment has all along hidden this politics of the thingly. It hid the Greek politics of the thingly just as well as it hides ours.

J. Donald Hughes, whose image of a temple surrounded by a decimated environment, points out that every ancient civilization of the Mediterranean Basin brought about the same result: the Greeks deforested Attica; the Phoenicians, Lebanon (whose cedars went to Israel to build temples not unlike the Athenians); the Latins, the Italian peninsula—and all with low-technology bronze axes.

We, of course, recognize some difference between this ancient rise of civilizations and their impact upon the earth and the modern equivalent. Brazil's rapid decimation of the rain forests is speedier because of chain saws and tree-cutting megamachines, hypocritically echoed in our own free cutting of the redwood forests north of here. High technologies amplify and magnify what a culture can do; and today we are aware of the threat of the greenhouse effect, whose gases, according to a very recent account in *Science,* are probably 20 to 25 percent homogenic in origin.

But now I have gone too far. My illustrative excesses may make it sound as if Heidegger was not one of the most important "fathers" of contemporary philosophy of technology and, worse, that those of us who are post-Heideggerian are even more negative on modern technology. Neither of these conclusions is my intent. Thus I must make two closing apologies.

To make my apologia to the more reverent Heideggerians amongst us, I recognize that one of the deep significances of his account of technology is to have shown it not be simply some collection of objects, but long pre-Kuhnian, a way of seeing, of revealing a world. And Heidegger is right, at least about the dominant way of Western seeing of nature as a resource well for human purposes. What is uniquely Western about this view lies mostly in its connection to the systematic mathematization of nature and the emptying of nature of any but inanimate and deanthropomorphized qualities. Unfortunately, many—indeed, most—prior cultures have also included components which frequently negatively impact upon environments.

A survey of so-called primitive cultures would reveal a startling ability—with very low technologies—to decimate environments. Slash and burn agriculture could subsist only in a rich, renewable forest so long as the human population remained low. The Pacific Islanders, who swept eastward centuries before Leif Eriksson got to the New World, made extinct species after species of wingless and winged birds in the easy search for food and fashion. And one could go on gloomily for some time. The exceptions of finely balanced minimalist societies such as the Inuit in the Arctic or the inland Aboriginals (for forty thousand years) in Central Australia, who left environments little affected, are also rare in previous times.

So, now my second apologia: it is not simply modern technology which threatens the environment. It is only the extent and the amplification of power which makes it global rather than regional. My negativity, rather, is addressed at cultures—which embed technologies but which heretofore have too often not been sensitive to the contextual and long-range effects of technologically enhanced human action.

Here, in conclusion, begins to emerge my reason for so thoroughly demythologizing romanticisms: there is no previous time to which we can return where the gathering of the fourfold was "right." The Greeks, the Romans, the Hebrews, our forefathers were not sufficiently concerned with our earth in any of their forms; nor were the peasants, who for all their preservation and building up of soils in Europe, were still the world's most populous per square mile area on the earth. They had the politics which allowed the same peasants who preserved the soil to harbor the most virulent anti-Semitism and nostalgic call for a purified homeland, which nurtured the Nazism of Heidegger's time. This ambiguity of preservation of the earth with destruction of a whole group of peoples is matched by another unnoted ambiguity in the Heideggerian corpus.

The networks—highways for the steel bridge, the electric grid for the hydroelectric plant—are also ambiguous. Because in the modern—and now I would say postmodern—world the network is what is beginning to make us aware of the displacement of our chauvinistic Eurocentrism which, to my mind, is linked with the romantic thesis not only in Heidegger but in our dominant views of technology, nature, and each other's cultures.

What is needed is not a rejection of the deep and essentially phenomenological insights into technology as a culturally embedded phenomenon with its different gestalt features, but a deepening and more complex appreciation of all of the facets of our technologically textured mode of life. And that includes and must include the explicit recognition of both the politics of our artifacts, and the demythologization of nostalgic and romantic views of previous times.

Instead, we need to develop a postmodern critique which, at this early juncture, is still in a bricolage stage. But out of our growing experience of cross-culturality we have begun to recognize that there is a plurality of cultures out there which threaten to decenter our past assumptions, and alongside which—but only alongside—we must reevaluate our past assumptions.

So, my demythologization of romanticism is also a critique. It is aimed at noting the freeing side of postmodern technological civilization and the opportunities which lie in its very networked ambiguity. Global pollution, the threat to the earth posed by our amplified powers, has also the promise of now seeing ourselves globally within a plurality of cultures. But none of these should, or ought to be, romanticized. Rather, our emerging but still primitive awareness of pluriculture should be taken only as a threshold for simultaneously freeing ourselves of a past fraught with too frequently bad ambiguities and opening ourselves to the uniqueness of a new world, equally ambiguous, but for the first time genuinely global.

The dramatic space shots of earth from the moon or a satellite are very un-Heideggerian precisely because they place earth at a distance from earth-as-ground. But they are also irreversibly part of the postmodern view of earth-as-globe with a very different sense of what constitutes our "home."

9

Feminism and the Philosophy of Science

Three decades after the English translation of Simone de Beauvoir's *The Second Sex* (1953) feminist critiques of science began to appear within the field of philosophy of science. It will be my aim in this essay to assess some of this early impact in three steps: (1) I shall first place feminist philosophy of science in the context of the explosion of philosophies of science in the last decade; (2) I shall examine what I take to be three of the most important goals of the feminist critiques—(a) the analysis of the "sex/gender system" as it operates within science, (b) the beginning development of gender-pluralized sciences and changes made evident, and (c) the arguments concerning overall transformations of science with respect to gender pluralization; and (3) I shall conclude with an evaluation of feminist philosophy of science within the contemporary field of that subdiscipline.

I. Feminist Philosophy of Science in Context

Institutionalized philosophy of science itself is relatively young in Anglophone circles. The North American Philosophy of Science Association was organized in 1934 and its history has been marked by three periods.

Period 1. From 1934 to 1962 one could identify philosophy of science as moving within a dominantly logical positivist and kin approach, which took science as a kind of theory generating system of hypothetical-deductive schemes. A. Pickering has characterized this approach as a "science-as-knowledge" approach—hereafter *SK*—and it could also be characterized as ahistorical, an-institutional, and apraxical.[1]

Period 2. Although far from alone in shifting to a much more "science-as-practice" perspective—hereafter *SP*—the publication of Thomas Kuhn's *The Structure of Scientific Revolutions* (1962) led a virtual palace revolution such that it is common to refer to "pre-Kuhnian" and "post-Kuhnian" philosophy of science. If Kuhn's approach—along with that of Paul Feyerabend, S. Toulmin, and others—could epistemologically be called "framework relative," what was much more important was the shift to perceiving science as an embodied and historical institution. From the sixties into the mid-seventies, much of the publication contestation in philosophy of science was a battle between *SK*, which gradually transformed into more modestly chastized pragmatic and lower altitude studies, and *SP*, which began to also appear in the early social studies of science and "STS" (science, technology, society) programs. One may note in passing that "old philosophies never die, but they do revise away."

Period 3. The third period might be characterized as a pluralist explosion in perspectival philosophies of science. This explosion was noted by Steve Fuller in a masterful survey of the field, "Philosophy of Science since Kuhn," for *Choice* at the end of 1989.[2] He dates publications from 1977 to 1989 as this pluralist explosion, to which I shall enter a few updated glosses. Fuller cites some eighty-three books published in this period—I will add a few he missed.

SK of course continued in its highly revised and usually more historicist and pragmatist chastised forms—only one book since 1977, claims Fuller, could still be identified with the positivist programs, Clark Glymour's *Theory and Evidence* (1980). *SP* had diversified outward to many social science and historical studies in philosophical modes as well.

Both of the traditions noted simultaneously became more science discipline specific, or, as I would add, more perspectival with respect to some praxis-set, and expanded outward to new areas beyond the almost exclusive obsession with physics in Period 1. These included the new analytic interests in AI (artificial intelligence), the cognitive sciences, and biology.

Perhaps the most powerful new perspective that gained attention was the emergence of the "social constructionists." This "school" of philosophers and social scientists cum philosophers began to look at the social construction of science, and in some degree began to interpret the "products" of scientific activity as "socially constructed" products—in principle no different from other socially constructed products. In passing I note only Andrew Pickering's *Constructing Quarks* (1984) and Bruno Latour's *Science in Action* (1987), often cited as the eighties' watershed book equivalent to the Kuhn's sixties' *Structure.* And although one may identify particular works as "social constructionist," social constructionist themes *run through many of the other developments of pluralist philosophies of science as well.*

Fuller also notes "the new wave" of serious and sustained "continental" philosophies of science which emerged in the eighties. These, drawing from the phenomenological, hermeneutic, and poststructuralist traditions—hereafter *PHS*—also begin to look at science as phenomenon—Fuller forefronts Patrick Heelan's *Space Perception and the Philosophy of Science* (1983) and Joseph Rouse's very Foucaultian *Knowledge and Power* (1987). But *PHS* themes, like those of social constructionism, also are pervasive amongst the new philosophies of science.

And, I will now gloss Fuller, there also emerged a loose school of philosophers who might be said to hold a thesis in some ways parallel to social constructionism, which might be called "technological constructionists" in a perspective upon science's *technological embodiments in instrumentation and experiment.* This school has now been highlighted in my *Instrumental Realism* (1991). It includes both continental types like Heelan and myself, but also more analytically oriented philosophers such as Ian Hacking and Robert Ackermann.

And—into the midst of this proliferation—there also begin to appear the first feminist oriented philosophies of science. Fuller cites, out of his eighty-three books, seven books by women, of which five could be clearly identified as feminist (I note in passing that he missed one important woman social constructionist, Karin Knorr-Cetina, *Science Observed* [1983] and two feminist collections by Ruth Bleier [1984 and 1986]). The most important of these—and I concur with Fuller—are Evelyn Fox Keller, *Reflections on Gender and Science* (1985); Sandra Harding, *The Science Question in Feminism* (1986); and Donna Haraway, *Primate Visions: Gender, Race, and Nature in the World of Modern Science* (1989). Since 1989, however, this genre has grown. Harding has, for example, published *Whose Science? Whose Knowledge?* (1991), showing even more momentum. The importance of these books, both in the discussion they have stimulated and in the issues surrounding the focal perspectives, outweighs the still relatively small number of the total production. (One may note in passing, however, that the percentage of books by women in this field *approximately reflects the gender ratios at senior levels in the actual field.*)

Before turning to the specific programs of feminist philosophies of science, I want to anticipate and foreclose several lines of argument surrounding this development. Much argument still echoes—as it did at the rise of Period 2 and *SP* interpretations—the retreat strategies of *SK* traditions. These arguments are those which wish to discredit or invalidate feminist perspectives as "philosophy"—but, interestingly, those arguments do not apply specifically to feminist versions alone, but to all the newly emergent *SP* traditions. While I shall note some of these in passing, I also want to note that (1) the battle is in one respect over, since the battlefield already contains in great strength all the newcomers; and (2) more interestingly, by the very presence of this diverse set of forces, *SK* itself in retrospect *must now appear as a special perspective* upon science.

There is one more significant observation which is relevant prior to turning explicitly to the feminist programs. Regarding *which* science plays the most paradigmatic role as a science, there can be little doubt that virtually all of Period 1 concentrates on *physics* as that science. But if physics is paradigm for *SK*, it remains central for much of *SP* as well, and, for most males in Period 3, it still retains a central place—Pickering's *Quarks*, Heelan's *Space Perception*, and, amongst the experiment people, Peter Galison's *How Experiments End*, all focus on physics. Amongst the feminist critics, however, there is a perceptible shift of attention to the biological sciences. This is, in part, expectable since actual gender distribution in the biological sciences is much better than in the physical sciences, but it may also be interesting to note that in the contemporary world there is also a strong shift of popular attention to precisely the spectrum which runs from sociobiology to biotechnology.

I note this because what is common now to virtually all contemporary philosophy of science is a kind of thematic perspectivalism. Such perspectivalism is often—although not exclusively—nonfoundationalist and complementarian in approach. These approaches are also frequently what I shall call "postmodern" or even "eclectic" with respect to the use of "theory-families." A theory family is some theory, a critical perspective, and interpretive context—hereafter *TCI*. But now I shall turn more specifically to the feminist programs for analysis and critique.

II. Three Feminist Programs

If feminist philosophies of science have a "strong program" it is the identification of, analysis of, and pervasive expansion of the sex/gender phenomenon as operational within science-as-institution. This makes the feminist programs, not unlike others noted in the postmodern context, deeply perspectival. In this feminist approaches are congruent with other thematic approaches to science.

Moreover, feminist theory relates to and develops from a range of *TCI*s already familiar in the contemporary world of the "new scholarships," including the neo-Marxian, poststructuralist, Foucaultian, and phenomenological-hermeneutic strands. While all these *TCI* approaches are new to philosophy of science, there is, in much feminist writing, also a highlighting of the neo-Freudian strands of interpretation. In this respect, feminist theory, as applied to science, is distinctly cognate to the spectrum of postmodern tendencies.

A. Program 1: The Sex/Gender System

Evelyn Fox Keller's *Reflections on Gender and Science* is the first to appear and to announce the foregrounded theme concerning the sex/gender system in science. She opens her study by clearly and explicitly placing this study in the *SP* and Period 3 characterization of science-as-institution:

> The essays in this book are premised on the recognition that both gender and science are socially constructed categories. Science is the name we give to a set of practices and a body of knowledge delineated by a community, not simply defined by the exigencies of logical proof and experimental verification. Similarly, masculine and feminine are categories defined by a culture, not by biological necessity.[3]

The question will be: how broadly and deeply pervasive *within science* is what is taken as a socially constructed sex/gender system? And, philosophically implied, if such a system goes both to the boundaries and the bottom, what does this imply for any notion of "scientific truth"? These are worries which feminist philosophies of science have, and share with other pluralist approaches to science-as-institution.

One of the deep sex/gender-knowledge associations which Fox Keller sets out to show is that there is a socially constructed identification of impersonal, objective, and value-neutral notions of truth with the masculine, and that, in a binary system this makes the contraries feminine. She amusingly cites Mary Ellman, "Faced with the charge that 'women always get personal,' Ellman counters: 'I'd say, men always get impersonal. If you hurt their feelings, they make Boyle's law out of it.' "[4] But, then, Keller goes on to add, "The fact that Boyle's law is not wrong must, however, not be forgotten."[5]

Feminists *as scientists* note that science and its accomplishments will not simply go away, even if they have been deeply flawed by the extremities of the sex/gender binarity. And thus the question of what is essential to the nature of reliable and intersubjective knowledge of the natural world remains within the context of the feminist critique.

If science is constituted within the sex/gender system, how broadly and deeply does this system cut? At the demographic level there is little argument: the "higher" in the corporate system—above lab technicians and the secretariat—and the "harder" the science—especially physics and its engineering cognates—the answer is clear. Natural science is contemporarily male dominant. Physics and engineering (using North American statistics) remain approximately only 4 percent female. Then, going down toward the ever "softer" sciences, one reaches biology, very broadly defined, which does attain up to 24 percent women.

The social sciences—often discounted as science by natural science peers—rise to some 35 percent, with nonscience humanities in some fields roughly reflecting the overall demographic pattern of a 50/50 ratio. Philosophy, incidentally, remains the "physics" of the humanities with now approximately 18 percent women.

Much early feminist activity went into sociopolitical programming which sought to enact affirmative action and educational programs designed to change these ratios—but with a few isolated exceptions, most of these programs have failed to make major demographic changes. Indeed, one could argue, in the traditional "harder" sciences, in the light of much Anglophonic country failure in recruiting ever larger cadres for Big Science, most of the recruitment successes have been among Asian and Indian males, i.e., males from cultures even more "patriarchal" than Anglophonia. The weakest statement which could be made at this level is that science-as-institution is imbalanced with respect to gender pluralization!

At the level of science's self-interpretation, however, the first defense will be one which maintains a strong distinction between ideal "gender-neutrality" and scientific construction. A popular rhetorical defense is the use of a counterexample—could there be a "feminist mathematics"? a "feminist physics"? Of course, it is assumed that such a notion is counterintuitive.

But, as I shall shortly show, in the light of some recent and significant social changes within isolated sciences, it may no longer be counterintuitive to think a "feminist primatology" or even a "feminist biology."

Fox Keller's program, however, is to show that the sex/gender system (a semiotics, if you will) is one which pervades the very history of truth itself. Her first three-chapter sequence on three moments of the history of science/philosophy are to date the most clear and succinct outlining this program.

The first of these moments is with Plato. She notes that not only does Plato explicitly model knowledge on sexuality, but on a particular type of sexuality—*idealized homoerotic sexuality*. Plato, she argues, not only first associates knowledge with sexuality, but is the first to develop a particular model of that sex/knowledge. It is both *nonheterosexual* in form—"Heterosexual desire does not contribute to transcendence, for Plato, precisely because of its association with physical procreation."[6]—and, ultimately, even in its preferred homoerotic form, is itself idealized and *nonphysical*.

Platonic love is *dis*embodied. But here I shall gloss Fox Keller: there are actually two moves or levels here. The first is a selective factor, the *choice* of a particular physical sexual model, male homosexuality, and the second, a move into typical Greek "transcendence" or disembodiment. The dualism of body and mind which pervades Greek thought, and which takes so many variants from Plato to Descartes to contemporary analytic philosophy, also now is shown to be a peculiar sex/gender variant. What is most powerful about Fox Keller's analysis is the argument that at this highest level there remains the idealized sex/gender choice.

But the transcendent or disembodied "ideal homoerotic" knowledge remains also "Platonic" in its strictly *contemplative* telos. Plato cannot, by himself, originate any but a purely theoretical or contemplative science. It is not accidental that *mathematics* should be the highest "science" correlate to the higher forms of the Good, the True, and the Beautiful. Plato is, at most, one side of Western science.

Fox Keller's second moment is that of Bacon. And, again, there is an explicit use of sexual metaphor—this time *heterosexual*. The model is both heterosexual and *aggressive*. If knowledge *is power*, its power is metaphorically like that of the (active) male enacting desire upon the (passive) female. Man must, in knowledge-is-power, master nature, which is still feminine. Keller says of Bacon, "It is Nature herself who is to be the bride, who requires taming, shaping, and subduing by the scientific mind."[7] These heterosexual, aggressive metaphors were to become the tradition of rhetoric in science to the very present. Fox Keller, quoting a contemporary, writes: "I liked to follow the workings of another mind through these minute, teasing investigations to see a relentless observer get hold of Nature and squeeze her until the sweat broke out all over her and her sphincters loosened."[8] (As tempting as this may be, I shall not comment upon the act to be committed there!) Harding documents the same kind of rhetoric in Nobel Prize acceptance speeches as recently as the early eighties. The rhetoric, insofar as it reveals a social sense to science, is hardly gender neutral.

But the ultimate retreat is always to the neutral, the transcendent, the apersonal, which characterizes, not the autobiographical, but the technical language of science. Here the third moment in Fox Keller's sex/knowledge history comes to the fore. For this she turns to the period of the institutionalization of modern science in the seventeenth century. Again, in a succinct and interesting way, Fox Keller shows the emergence of the dominance of the *mechanical* metaphors which also retain their power in contemporary science. The correlate of a neutral, disembodied, apersonal language, is a mechanical, unalive, passive nature. The victories of this crucial period are of the mechanists over the hermetic traditions, and of the scientific suppression of "female excess," presumably embodied in witchcraft.

At this point I wish to depart from Fox Keller for a moment. The framework of interpretation she uses is one strongly informed by variants on psychoanalytic theory—this is a framework employed in some degree by many feminists. I, personally, think this is a mistake. And I would argue that a much stronger case for coming to similar conclusions about the sex/gender role in the very precincts of science's language could come from hermeneutics and contemporary metaphor theory.

What Fox Keller's history of the rise of the Royal Society and the battle which "officially" chose a mechanical set of metaphors could show is that at the heart of science itself there is always a guiding set of metaphors which reveal and conceal in certain, rather than other ways. Much work has been done here to show how metaphoricity operates both positively and negatively, but in every case casting a certain aura or trajectory to the process. A second move would be then to show that there are "good" or approved uses of such guiding metaphors—mechanical ones at the rise of modern science—and "bad" or disputed ones, which today are the sex/gender ones which are resisted in the retreatist strategies I have noted. But, *both are and can be operational and this is what the analysis could and does, indirectly, show.*

In short, what we have here is a general argument about how metaphor does or does not operate within science. If—as most historians and *SP* philosophers of science hold—metaphors *do* form a kind of macroperceptual grid, then, once having shown the pervasiveness of sex/gender system metaphors they will have been shown to be just as important in some dimensions of science as the mechanical ones were in other ways.

Now, I wish to gloss Fox Keller. I suggest that a fourth moment could and should have been added to adumbrate the movement from contemplative, to knowledge-is-power, to the rise of a "resource well" of "mechanical Nature." That moment would be Foucaultian and reverse the knowledge-power equation of Bacon to the more familiar late twentieth-century "power-constitutes-knowledge," which more and more can be seen to be the social constitution of a corporate modeled Big Science, and which again is distinct from the three forms noted here. In this glossed moment, the aggressive language returns in the form of science entrepreneurship.

But if the sex/gender and sex/knowledge relationship is brought into prominent view, what now? Interestingly, I find two countertactics are the most popular. The first is to *dissociate* science from its rhetoric. This move, we saw above, returns us to the old idealization movement into "Science" as pure knowledge. But the other—which I have actually heard on many occasions on the part of scientists more than philosophers—is, "so what?" Here is an implicit admission that "we knew that all along" and then simply *affirm it.* That is the rhetoric which can and does occur in the power centers of science. It was also evidenced by that arch-masculinist, James Watson, of double helix fame. Harding quotes his response when he complains that the Federal Government gets too much advice from the "harder" sciences of physics and chemistry. Then for biology, he moans, "The person in charge of biology is either a woman or unimportant. They had to put a woman someplace."[9]

In short, what if the retreat strategy of the philosophers were given up, and science simply asserted itself as explicitly masculinist? I suggest that it would then move as easily as it does with its other Big counterparts, Industry, Multi-Nationals, the Military, or, metaphorically, the Locker Room. But is not this where it already is? And, if so, have the feminists done more than to make explicit what has been implicit all along?

If it is still counterintuitive to speak of a "feminist physics," imagine a fantasy variation: what if physics became, overnight, a totally feminist enterprise? I shall not imagine what shape it might take internally, but taking just some of the values espoused, which might include more maternalistic and care concerns, which might be more decentralized and democratically directed, etc., etc.—what would happen with respect to the larger interconnection with Industry, the Multi-Nationals, the mega-Universities? My point is this: the problem is much larger than even science per se, and implies that to change science, one must change a world.

In short, I am suggesting that insofar as power-constitutes-knowledge in the present sense, the battle will not be won by making concepts clear, or by seeing their contextualization, no matter how accurately, in their sex/knowledge structures. Thus, we must move beyond intellectual therapy.

B. Program 2: Gender Pluralization in Science

One of the virtues of a "science-as-practice" perspective is that actual social and historical examples may be taken as test cases with respect to theory development. What would be needed as a kind of test case regarding feminism in science would be to locate precisely such change with respect to women's impact upon science. This is precisely what Donna Haraway, biologist turned historian/philosopher, does in *Primate Visions*. I regard this book as the most original and positive of the trio being dealt with here.

Primate Visions is a history of often dramatic changes in the history and social constitution of primatology, an admittedly sociobiological and interdisciplinary science. But it is the science which may also be the one most changed by the rise to power and perhaps even dominance by women scientists. Haraway's book is a history—but also a highly postmodernist and hermeneutically oriented reading—of that science.

In very briefest form, women primatologists—in spite of the paternalistic patronage of males in the field and the heavily romanticized publicity practices of *The National Geographic*—became the leading researchers in the field. All will recognize the best publicized of these: Jane Goodall and Dian Fossey. But the narrower and perhaps more telling paradigm-shifting work belongs to lesser known figures, among them, Sarah Blaffer Hrdy.

I shall later return to comment on the remarkably different research style that also emerge in this history—a research style which emphasizes not so much the aggressive sphincter loosing noted, but a close, detailed, daily, and empathetic approach to one's subject, not incidentally missed by Fox Keller's work on Barbara McClintock's work on genes, and clearly exemplified by all the field primatologists who came to prominence. But, instead, I shall concentrate on the paradigm shifts that eventuated from this work.

To illustrate what has today become a major biological paradigm shift, I shall revert to a single example which exemplifies the now widespread change of not merely primate, but much mate selection modeling noted recently in *Science*. From Darwin's revolution constituting evolutionary interpretations of animal life, there remained a certain Victorian caste to male/female animal roles in selective mating. Hrdy called this the "myth of the coy female." Put oversimply, the "active" male is simply opportunistic, casting his ever active phallic energies at any female whatsoever; whereas the "coy" or "passive" female refuses *all but her selected mate* and then, presumably in the higher animals, tends to be monogamous. This Darwinian picture does credit *female choice* with a selective role.

Hrdy's early work with lemurs, however, showed a very different pattern—one which today stands much closer to the contemporary paradigm for mate selectivity. Female lemurs, while sometimes appearing to favor mating with a dominant male, often were very promiscuous; even promiscuous on the sly, with multiple partners, even whole gangs of not yet dominate bachelors. In short, worried Hrdy, the female of the species was herself aggressive, promiscuous, and appeared less like a Victorian lady than a sixties' and seventies' feminist libertarian! (The time at which her research was being performed.) As it happened, this set of observations also meshed nicely with the rise of the "jealous gene" theory also contemporarily popular, and Hrdy could see such behavior as a way of sexually persuading males not to kill offspring in the chance that the victim might be one of their own, and even of soliciting childcare favors from former lovers.

Today, such animal behavior has been observed and come to the fore, not only with primates, but with birds, many mammals, etc. (My male Stony Brook colleague in animal psychology has gone on to show primate deception practices which show that loud groans and moans in mating are often reserved for the dominate males, but silent and hidden mating with younger "turks" goes on quite commonly beneath the foliage!)

One can here detect something "all too human" and even "restoration comedy–like" in these studies—but my point is not amusement, but an illustration of how a whole change of paradigm model came about by and through the gender pluralization of a single discipline, and then spread throughout wider sociobiological sciences. In part, it was due simply to asking different questions and changing focus—for example, much of the transformation came from a shift from analogs of leadership power structures amongst baboons, an earlier interest of male primatologists, to infant- and childcare strategies amongst female primates, interests of female primatologists. Primatology was one of the first, and remains an exemplary instance of a gender-pluralized science *and it is a science in which the models, the questions, the perspectives, and even the research methods have been changed as a result.*

I want now to return briefly to some remarks about research styles. Two of the great strengths of Haraway's book revolve around how she shows the styles involved in both setting up experiments and, then, the public portrayal and propagation of these. Her chosen theory families utilize both a hermeneutic and narrativological sensitivity to the "stories" in which science casts itself.

In primatology the gender contrast is often striking—Haraway's histories of the "Apes in Space" experiments and "Metaphors into Hardware," the first regarding the use of experimental animals in early space exploration and the second the infamous surrogate mechanical mothers in isolation-depression studies, are often close to bloodcurdling. In contrast—and in spite of the excesses of someone as idiosyncratic as Fossey—the empathy bonding often displayed amongst female primatological researchers and the metaphors which emerge therefrom is often stark.

The real question, however, is one concerning whether or not new and scientifically constituted knowledge arises in spite of the dramatic shift of interpretive context. I would hold that it does, and that it does so in both a somewhat surprising way and in parallel to some of the favored "best science" in the still masculinist strongholds. Today one cannot doubt the importance of Nobelist McClintock's work on genes. And in the shift noted in primatology, the knowledge gains have been vast and include—in addition to the mating practice shifts—a demythologization of any "noble savage" notion about primates (Goodall's discoveries about tool use, cannibalism, infanticides, and carnivorism amongst chimpanzees) in spite of both the traditions of the discipline and utopian hopes to the contrary. Goodall, indeed, had to place her own child in a cage while the chimps performed their cultural feats "free in the wild."

Haraway—in a striking parallelism with Peter Galison's similar investigation of two different instrumental traditions in physics, the discovery of neutral currents by both the accelerator people and the bubble chamber people in high energy physics, where the end of the experiment is "we can't make these effects go away"—concludes, likewise, that through the new feminist perspectives, certain effects do not go away.

I would also suggest that this parallelism points to an interesting and deep ambiguity in scientific constitution. It at least points to no specific *necessity* for the maintenance of the distance, contemplative, or even transcendent metaphors which have often contexted both classical and modern science. Could it be that a multiperspectival, eclectic *postmodern* science is possible?

C. Program 3: Changing Science Itself

With the third program, a strong normative element enters the field. Early *SK* philosophy of science could largely be characterized as simply impressed with the attainments of science and devoted largely to analyzing these. But a similar motivation, although on a broader front, lies within much early *SP* philosophy of science as well.

By the mid-sixties, and accelerating into the eighties, there also arose a feeling amongst many of the newcomers—particularly from *PHS*, philosophy of technology, and now, feminists—*that there was something also wrong with modern science.* Here the feminists join a small, but growing group within but not coextensive with *SP* perspectives.

If Keller was successful in showing the breadth and depth of sex/gender factors in the ideological operations of science, and if Haraway shows something of the historical/empirical possibilities in a gender-pluralized science, what of a possible gender pluralization of science (*Uberhaupt*)? This is the third, and normative program. What would remain of science's contributions, of the normative shape of science, were it to be gender pluralized? In this final program analysis I want to turn briefly to the feminist projections concerning the future of science.

All three of our highlighted authors acknowledge that there has been a movement among feminists toward science from initial reformist to more radical attitudes, particularly in the light of the already noted failures to gender pluralize virtually all "hard" science. Keller notes, "one strand of the radical feminist critique goes on from the hypothesis of deep-rooted androcentrism in science either to reject science altogether, or demand that it be replaced—in toto—by a radically different science."[10] But she goes on to recognize that she has learned too much as a scientist to give up on its knowledge gathering successes, and sees the danger of women simply remaining "outside" in the second.

Her own conservative solution is to underline and encourage high examples of successful role models—such as McClintock—in whom she sees embodied a different set of cognitive and perceptual scientific models of investigation. These are those which have a greater "feeling for the organism" such as we have already noted. This would be a reclamation within science, but it would also need to depend upon the greater gender pluralization of science-as-institution. Such a solution is in keeping with the implicit individualism of a psychoanalytic *TCI* upon which Keller depends, but it also reveals the weakness of precisely such a *TCI* for social and institutional change.

Haraway, as usual among the three, is the most imaginative and concludes *Primate Visions* with a chapter on what could be called "fictional multiple worlds." The chapter, "Science Fiction, Fictions of Science," draws on her somewhat postmodernist *TCI* and recommends something like a cross-gender, cross-cultural imagination to adumbrate (within primatology) the possibilities which also would break the present bonds of patriarchy.

So far I have not highlighted Harding, who is in fact the most philosophical and philosophically self-aware of the three feminist philosophers of science. Her normative recommendation is, in fact, the most radical of the three. She recognizes the present science hierarchy—espoused by the early *SK* philosophers of science and instituted in the present sociopolitical milieu. It is a hierarchy in which physics occupies the pinnacle, followed downward by ever less or "softer" quantitative sciences, and having at bottom the qualitatively focused ones including the social sciences.

By showing the implicit political and social *and gender* biases which pervade even the pinnacle, Harding recommends *an inversion of the science-value pyramid, i.e., a return to and inversion of the positivist "unity of science" program.* Her recommendation is one which turns out to be virtually identical to much *PHS* or continental conception regarding the sciences. She argues, normatively, that it should be the moral and political sciences which top the pyramid, with the narrower physical sciences as special sciences at the *bottom* of the pyramid.

The problems *within science* are, for Harding, exactly the most pressing problems within society at large—sexism, racism, classism, etc.—and these are hidden by the very existence of the implicit value hierarchy concerning the sciences, and the vestigially remaining notions of value-neutrality within science.

> Whatever the moral and political values and interests responsible for selecting problems, theories, methods and interpretations of research, they reappear at the other end of inquiry as the moral and political universe that science projects as natural and thereby helps legitimate. In this respect, science is no different from . . . computers: "junk in, junk out." It is within moral and political discourses that we should expect to find paradigms of rational discourse, not in scientific discourses claiming to have disavowed morals and politics.[11]

By asserting the priority of moral and political over epistemological and scientific discourses, Harding also holds that we must see science as a particular episode in modern history, which must be transformed into a postmodern episode that transcends the modern narrowness. Interestingly, in developing such a perspective, she also is following the same trajectory that is also followed by both very contemporary philosophy in its postfoundationalism, and by the pluralist philosophies of science.

But, will or could any of the above actually happen?

III. Conclusion: Feminist Philosophy of Science and the Future of Science

I shall conclude with a parabolic, irreverent detour. If it is fashionable in the light of general, and now philosophy of science, feminism to have forefronted considerations of the sex/gender system, it may be noted that one clearly "socially constructed" variable on these roles has not achieved much contemporary and secular discussion. Familiar is hetero-, homo-, bi-, and even trans-sexuality in the discussion—but what about *celibacy*? And, we have a 950-year-long institutional example of this sex/gender variant—the rule for a celibate clergy in the Latin Church.

In its history, clerical celibacy was anomalous. There were no precedents from the Judaic side—the Orthodox rabbinate remains heterosexual, monogamous, family prolific, and strong gender role differentiated. On the Greco-Hellenic-Roman side there were some precedents including mystery religions, vestal virgins, and *some early philosophies* including elements of Pythagoreanism, Epicureanism, and the Platonic recommendations noted by Keller. All of these systems devalued the body, were dualistic, and elevated a disembodied form of ideality to the top of their value hierarchies. This was carried over into the association of celibacy and the clergy in post-Judaic Latin Christendom.

But clerical celibacy did not enter serious discussion until the fourth century—it was rejected by the Nicean Council in 325—and did not become regulated until the Councils of 1123 and 1139. Now, it might be expected that such an anomalous sexual role choice might cause problems between theory and practice! It did, as period after period of Church history showed with regular episodes of practical concubinage, homosexuality, and the famous diaries of monastic celibates concerning autosexuality (masturbation). Indeed, the open disregard of rules by the Renaissance papacies and higher clergy was one factor in the Reformation, which broke the hegemony of Rome and historically reduced the Church (with a capital) to pluralized churches (lower case). And, with this episode, clerical celibacy became a clear minoritarian "choice" even within Christendom.

This irreverent parable, however, contains many interesting parallelisms with precisely the history of the philosophy of science I have been tracing. I cannot now flesh all these out, but want to suggest that one such analog is related to the shift from *SK* to *SP*. When an ideality—pure and objective science viewed only epistemologically—contrasts too strongly with practice—the technoscience involvements with the whole "military-industrial complex" and the now crisis with a global environment—I suggest that the shift of perspective between the ideal "Church" projected in the Renaissance and its practices, becomes almost inevitable. And I would argue that the Reformation, Renaissance humanism, and other *SP* equivalents of the time, were perspectives which took shape in the light of this divergence.

In short, I am suggesting that the proliferation of pluralist philosophies of science—many with strong normative programs—are the late twentieth-century response to a much broader shifted perspective on science overall. Feminists are not alone within this field—but they have hit home with respect to a nondeniable bias within the very heart of science-as-practice. I doubt that it will get off free.

If the eighties, and now nineties, are in any way the equivalents to the shift of perspective of the Reformation, perhaps the deconstruction of science in its presently organized hierarchical mode, is also being heralded?

EPILOGUE: ''THEORY''

Although it really was the era of the sixties which began to introduce what later became known as "theory" in the jargon of the literarily oriented disciplines, today the battlelines between academic "revolutionaries" and "traditionalists" often revolve around a positive or a negative take on "theory."

Marks of the earlier revolutions include those already located in the coming of the new philosophies of science, often simply dubbed "post-Kuhnian." Similarly, the early sixties saw the first intrusions of European-styled philosophies begin their challenge of the then dominant analytic or Anglo-American ones. Early feminism also made its appearance, only later to take more clearly academic shape in actual programs or departments. To these have now been added a whole spectrum of ethnic and cultural studies programs and the ongoing discussions and debates over the curriculum.

In this final section, I take a look at the academic world in precisely the above context. Within the humanities the whole variety of "new scholarships" or "theory" contest for a place within departments and the curriculum. But the humanities are so often dwarfed and overshadowed by the technoscience disciplines that in one respect it is a wonder that the tempests within the humanities could reach such a national level of attention. Yet, in a rather ironic way, the challenges of feminism or multiculturalism are often taken at least as seriously—maybe today even more seriously—than the older seeming threats of socialist ideology. In part this is due to the prominence during the last decade plus of the influence of the far right which so successfully penetrated governmental precincts. The politicization, far from the minoritarian calls from the new arrivals alone, took shape in the appointments and directions of the NEH and NEA, whose directors under the previous two presidencies were carefully screened for conservative politically correct views.

Today there are signs that some of this dominance is beginning to unravel, particularly in the light of the collapse of the old demons of socialism and communism. But all of this would be short-sighted. The academy and the curriculum are changing and will change. The very populations of our universities have changed, as has the nature of our emigrants, and with this change there will be a "reinvention" of our history in the light of our present and future.

Even deeper, however, are the intellectual changes within our awareness, which came about by virtue of the very mode of investigation honed to a high art in the West, the mode of scientific-instrumentally mediated investigation which has changed our past(s) already. And while I do not really pretend to know what the humanities will actually look like in the year 2000, this final chapter does try to elevate some of the subterranean aspects of the current crises in the humanities.

10

Humanities in the Twenty-first Century

The previous chapters arose out of the late eighties and early nineties, but we are now rapidly approaching the millennium, the year 2000. I have already alluded to the ways in which our symbolic "discoverer," Columbus, has been reinvested at each centennial. Millennial inventions must thereby exceed even the former. Yet, two thousand years does not seem to carry with it either the sense of doom or the Day of Judgment, or the utopian Day of Resurrection that earlier millennia in our era seemed to hold.

If it is a time of revived fundamentalisms, it is one of fundamentalisms which far exceed those of merely Euro-American character, with Islamic, Hindu, and indigenous forms only once again underlining the essential *pluricultural* dimension to everything postmodern.

Here, however, I wish to focus on the academy itself, particularly on what once was its liberal arts curriculum and, within that, on the current and future fate of the humanities. One of the first things that becomes evident is that the humanities today play whatever role they play in the shadow of the "big" technoscience disciplines which themselves have grown differently in this latter part of the twentieth century.

I. A Close-Up View; or, From inside the Academy

John Compton, well-known philosopher and son of Arthur Compton, the Nobel Laureate for Physics, once noted that the basic work his father had done for the prize was done from handmade tools in the basement of the Physics Building. That was as recently as the 1920s, but today it is virtually impossible to imagine physics operating without mega-instrumentation.

The latest tool—over which there has been much federal argument—is the super collider, of which the still undercosted budget already exceeds eight billion dollars. Biology, the latest science to become "big" began with a single billion dollar project, the Human Genome Project, which today is already estimated now at three billion. And once beyond such "pure" science and into the various marriages with engineering, industry, and defense, even the above projects are dwarfed. The space station is (under)estimated to come in at fifty billion and SDI or "Star Wars" (which does not die, it merely transmutates) once was budgeted for one *trillion* dollars.

Or, if one looks at the funding agencies, which relate most directly to primarily our research universities, it is easy to make a more direct comparison with the National Science Foundation carrying over a two billion dollar budget which has doubled within the time of the Reagan-Bush era, and that of the National Endowment for the Humanities which has remained basically stable at approximately one hundred and forty million. The NSF "education program" division carries more than this and supports proscience advertisements even on children's television! Similarly, the National Endowment for the Arts has remained constant, while battling off the right wing which decries its support of presumed "obscenity" (in contrast to the implicit *obscenity* in the megabillion dollar level of support for purely destructive technologies!).

If this is a national perspective, how does it translate into the academy? First, it does so in comparative budgets: on my campus, a "class one research university" (a category almost solely determined by exceeding 133 million research dollars), this means that the science budgets are not only the largest dollar items within the university, but in their now infinite hunger for more "leveraging" and "matching funds" become indirectly the largest factors over all academic policy.

All of this indirectly translates into where the curriculum will be supported and where not. But even prior to the academic curricular battles which mark the nineties, it is interesting to note that external to the academy and at the elementary and secondary levels, technoscience education plays a highly visible and powerful role.

One of the major problems facing technoscience education is its failure to successfully recruit the numbers needed for the future of its institutional needs. This is not from lack of trying or publicity. The destruction of the *Challenger* revealed an interesting pathology: the school teacher killed, Christa McAuliffe, had been made into a kind of elementary child's pioneer role symbol through the pervasive and slick NASA materials in virtually all elementary schools. When this programmed heroine was tragically killed it became an event far out of proportion to other events which were both farther away and of much larger scope.

Similarly, the science documentary, excellently and dramatically crafted, saturates much educational and public channel television. What child today is ignorant of black holes, of the behavior of sharks, of the spectrum of the most esoteric animals in Australia, or of the decimation of the rain forests or whales? The science documentary is some of the best drama available on the tube.

Yet, in spite of massive funding, media saturation, and the dominance of science administrations, particularly in large and public universities, there is an enigma—the system is failing to deliver the bodies needed for the growth and nurture of Big Science. The most obvious failure—sometimes soft-pedaled, except in technical journals because of worries over state and national funding backlashes—occurs at the graduate levels of engineering education. It is widely known, inside the academy, that over 50 percent of all graduate engineering students in the U.S. today are foreign students (at my university that number is 83 percent).

India, for example, is a primary exporter of graduate technological talent. The most selective of schools of that system are its IITs (Indian Institutes of Technology). For Bombay, over seven thousand students take the entrance exams, with only fifteen hundred making it. Then, after a further weeding out over the years of the B.S. degree, some fifty each graduated from the disciplines of computer science, materials science, and electrical engineering—and all fifty from each graduating class came to the U.S. for graduate education, with an estimate that only 10 percent would ever return to India. This "brain drain" is deliberate on both sides: many—probably most—Indian students see the degree as the ticket to the wealthy West; and the engineering schools in the West see it as a necessary feeder to keep the world of engineering healthy.

The second national phenomenon, much more widely publicized, is the relatively poor performance in mathematics and science by U.S. students. In a recent survey of seventeen countries concerning fourteen-year-olds in math and science education, the U.S. came in fourteenth. We beat Hong Kong and the Philippines, tied Thailand and Singapore, but came in well behind every European and other Asian country and very far behind Canada. Again, a small comparison between the values of technological society and the humanities and arts: in the fifties, with the success of Sputnik, a "missile crisis" was announced and a first round of national effort went into the "missile gap." Would it not sound odd to you, were I to proclaim a "ballet gap" wider than the "missile gap"? And, yet, that is the case—for in spite of our problems maintaining Big Science, the Russians are really much further ahead of us in ballet than in space technology.

Within the academy, the humanities remain the poor cousins. They are needed for certain purposes: entertainment, relaxation, leisure time stimulus, and—increasingly, for another purpose which I shall address—*to convey a certain appreciation for a tradition, quite specifically defined.*

This, then, is a view from close up, from what is seen to occur within the university. The humanities and arts take their place within the context of a more dominant cultural force, technologically defined and valued civilization, that form of civilization which, even if sighted as a trajectory in early modern times, only came to clear dominance in the nineteenth and twentieth centuries.

II. The Wider Angle, Global Impact

The technological civilization just glimpsed close up also has had a recent historical impact on the world as a whole. For one of the effects of recent science and technology—the "global village effect"—has been to shrink the time and space of the globe between all peoples. This effect, however, is also a condition for a series of changes facing us for the twenty-first century. Let us briefly examine some of the particulars of this effect/condition.

1. One aspect of global shrinkage is to have finally expanded the previous recent centuries' voyages of discovery into every nook of the globe and thus to have uncovered (probably) every remaining "primitive

culture." The Tasaday, discovered as recently as the seventies, touted as an undiscovered tribe in the Philippines, were later found to have already had some contact with the outside world. These vanishing worlds, we have discovered, are fragile; and once the relative territorial and cultural isolation is broken, the culture faces disintegration in a form not unlike the decline in marsh bird species once the marshes are drained. For the very contact often comes through the destruction of the ecological niche necessary for the existence of a primitive culture.

2. And although the remaining primitive cultures often begin early signs of disintegration shortly after contact, there is no cultural group which today can avoid a societal decision about what to do about technological culture. Whether that response is an embrace and desire for modern science and technology, or opposition and rejection, or an attempt to modify what and how it is adapted, no larger grouping is without this issue in the present.

In countries such as India, which is trying to learn and evolve its own autonomy in development, or South America, whose intellectuals see much of modern science-technology as a form of cultural oppression, or parts of Africa which see in technology a means to leap into the twenty-first century without technological evolution, some decision is being made.

And, no matter how the decision is being made, it cannot be avoided, and its consequences are not neutral. Perhaps this is most obvious in the interface between high technology and popular culture as focused in the phenomenon of television: Pakistan, not unlike many other Islamic countries, desires television—but under a strict set of controls, and without allowing certain forms of Western programming. Here, a major issue is a feared impact on women and their self-conception of role under the customs of purdah (the veil) and the home (women belong primarily to the private sector).

Twentieth-century technological culture is a culture-transforming force. It is, in some ways, radically acidic toward precisely those cultures which have had long, insulated, and tight-knit traditions.

3. But if the first two points indicate something of the impact of technological culture on traditional cultures, there is also another side to this effect which reverberates more directly upon the situation of the humanities toward the twenty-first century.

That is, the same means by which other cultures have been touched by largely Western originated and oriented technology, have become the means by which those cultures enter into *our* consciousness in a new way. This is the positive side of the global village effect. We are all aware that we can reach any sector of the globe by current air travel within a day, and we nightly touch many of the most diverse corners of the world of our more immediate neighbors. Ireland, Israel, South Africa, Afghanistan, face us almost daily in the news. But also in our entertainment we have become familiar with the Others through popular series such as "Shogun," "Jewel in the Crown," or through the ever-present documentary—who does not know about the seven-foot Masai and their mixture of milk and blood, or about the more ancient practices of the abandonment of the aged by the classical Eskimo culture?

Ours has become a world where the diversity of human cultures is commonplace. Our very histories have become interwoven so that we know of the similarities and historical connections between noodles, pasta, and lo mein going all the way back to Marco Polo.

4. Yet this cross-culturality is not merely a matter of curiosity—it is also something which ever more deeply is changing the face of America itself. If our historically dominant ties have been Euro-American, and our cultural lineage equally dominantly Judeo-Christian, it is not clear that this will be the case in the twenty-first century.

Our latest waves of immigrants have been Hispanic from the south (still within the Euro-American circle, of course) but increasingly from Islamic countries and Asia. In the now two decades I have been at Stony Brook, I have seen the Asian-American undergraduate population alone go from 3 percent to 15 percent.

Nor is the change of immigration pattern the only shift in the blending which makes us what we are. It is already the case that the dominant trade patterns are also changing from West to East. No one escapes the invasion of Japanese and Korean technology (the automobile, for example) in daily life, but this is simply the manifest sign of the coming Century of the Pacific Basin. And in the wake of trade follows the impact of culture.

The twenty-first century will be a century of shifted emphasis.

To this point, I have traced the outline of a context into which the current state of the humanities can be located. It is a context in which the dominant forces either are directly derived from technological civilization or are indirect effects and conditions which form the background for contemporary humanities disciplines. Thus, it is now time to turn more directly to those disciplines themselves to determine what trajectories can be seen with respect to the twenty-first century future. I begin with a look at what is currently expected of the humanities and their current health.

III. The Critique of the Humanities

In the face of these perceived changes, a critique of the humanities as a counterpart of the previously cited worries over science education has also arisen. Its opening bell was perhaps sounded by William Bennett, then head of the NEH prior to his becoming secretary of education.

In a report from a group studying humanities in higher education, but written in Bennett's own inimitable prose, *To Reclaim a Legacy*, the claim was made that our secondary and college-age students are woefully ignorant of the history and culture of Western civilization. They claimed that 50 percent of college students could get by without taking a foreign language, 72 percent without American literature or history, and 75 percent without European history, etc.[1]

What was assumed and stated was that a primary role for the college or university was that it "should recognize and accept its vital role as *conveyor of the accumulated wisdom of our civilization*."[2] And, it was equally clear that this set of core requirements which would be proposed as a solution to the critique should primarily be those associated with Euro-American history. What was needed was "an adequate education in the culture and civilization of which they [the students] are members."[3]

Since then, a chorus has arisen to play the same sounds, most notably the best-selling *The Closing of the American Mind* by Alan Bloom, but also echoed in a different way by E. D. Hirsch's *Cultural Literacy: What Every American Should Know*.

Who could possibly disagree with such a diagnosis? Just as in the sciences, these critiques point to a crisis in education which also embraces the humanities. But there are some notable differences in the shape this crisis takes in comparing the sciences to the humanities, differences which will become crucial to note later.

1. Whereas the knowledge deficit in the sciences, particularly math, is in many ways much more a problem of adequately teaching *methods* (how to think mathematically, or how to conduct experiments, how to attain objectivity and means by which data are to be gathered), in the humanities the focus is clearly much more on *content*, particularly historico-cultural content. It is: which books to read, which texts to know, which linear history to be familiar with. In short, one crisis is much more focused on the contemporary, the other, on the past. One is focused more on how-to, the other on what-about.

2. There is also a vast difference in the assumptions which form the background to these related crises. In the first, it is assumed that science and math education is, in a strong sense, *universal*, whereas the

focus of the Bennett/Bloom critique is clearly specific—it is *Western* history and culture which must be taught.

If these differences belong in any way to a single crisis in American education, it is not yet apparent how they interrelate. Indeed, I shall contend that there are subterranean linkages here which do in fact bring both the humanities and science education crises together.

IV. The Exploded Canon

On the surface, the critique of the humanities may thus first appear to be the counterpart to the previously alluded to failure of the sciences to deliver enough American youth with high enough qualifications to maintain the vast science-technology establishment in its present expansionist phase.

I do not wish to deny certain elements to that critique—but I shall locate the blame for this situation, not where the critics place it, but in a more general malaise. For I do not think any of the above symptoms are caused by simple curricular failures, nor can they be solved by simplistic solutions which would return us to a mythical time of prepluralistic American culture.

The social conditions that create the problem are reflected in two secondary phenomena which also can be seen today. For example, one of the reasons why engineering schools have a predominance of foreign *graduate* students turns out to be the simple fact that a B.S. in engineering today is sufficient to produce a quick-fix, high-salaried position for the American undergraduate. Similarly, the large number of surveys which show that the values of quick, high-income rewards are more highly valued than slower, lower-paying reflective disciplines (such as those which raise the question of the meaning of life), have motivated the even stronger turn of undergraduates to business and entrepreneurial programs that require less focused disciplinary training than the sciences; both point to a certain cultural mentality in which quick return takes top position. Indeed, still other surveys have indicated a certain uniformity to youthful dreams in the present: a Porsche in the garage, a six-figure salary before thirty years of age, and a life-style marked by brand-name, consumer objects of high quality.

Somehow, I do not believe a required curriculum of the Greek classics will remedy this situation. Nor do I think that a society which insists on paying and treating teachers with salaries considerably below those of sanitation workers, plumbers, and electricians, coupled with regulations and bureaucratic rules appropriate to the IRS or the State Government, is going to change this through any curricular reform. Our youth, unfortunately, are simply reflecting in typically impatient adolescent style, the very values which we have imprimatured in the presence in the White House since the Reagan-Bush era. Indeed, the only criticism I would offer such youth is that they have gone as extremely toward an embrace of current national values at the top as those in the sixties went in rejection of those values two decades ago.

Surfaces, however, are not what we are after. What lies much deeper in the debate about the humanities and their role is a crisis about a *canon*. We are most familiar with the concept of a canon with respect to the great religious traditions. In that context, a canon is an agreed-upon text or set of texts which are elevated to primary status to which and upon which other texts may refer in a secondary way.

We are also aware that in our variants on the Judeo-Christian traditions we have different canons. The Bible of the Jewish tradition is the Old Testament of the Christians, but its secondary texts are also the distinctive histories of the Talmud, the Haggadah and Halacha, and the Mishna. And even between the Catholic and Protestant variants there is some distinction between the roles or nonroles played by the Apocryphal Books. And, outside the Judeo-Christian religions of the Book, there is Islam with the Koran as a primary canonical text.

To anticipate some later points, it might be wise to indicate that the relatively clear sense of a canonical literature is actually unique precisely due to a certain cultural direction—that mentioned as the religions of the Books. Once one turns to non-Western religions, the notion of a canon becomes increasingly blurred.

What, then, is a canon within the humanities? Here, canon is obviously metaphorical, since there has never been any strong agreement on a single set of primary as compared to secondary texts. And yet, that is much of what motivates the current debates about core curricula: "Obviously"—the proponents argue—we should include Plato, Aristotle, Homer (and expand to other ancient Greeks), then Augustine, Aquinas (expand through the Middle Ages), then Locke, Hume, Rousseau (expand through the early moderns), etc.

But have we taken the notion of a canon as itself orthodox? And, if we have, has that canon—pardon the pun—*exploded?* I should like to suggest that in several ways there are today serious questions to be raised about the notion of a canon.

1. I have already indicated that the twentieth century is one

marked by the highest degree of cross-culturality of any century. No more than the fragile primitive cultures which today are vestigial artifacts of a once-isolated cultural ecology can contemporary culture simply ignore or dogmatically retreat into a monocultural heritage. It is simply provincial, or "Eurocentric," to assume that the great books and the great texts are those which belong to our linear heritage.

The truly educated person must place alongside the Bible and Homer and Plato the Rg. Veda, the Upanishads, the Book of the Dead, Lao Tzu, the Analects of Confucius, and a host of texts from great civilizations as ancient as or more ancient than ours. But how can we do this? The intercultural canon is simply too large to digest. How could each of us know everything? Must we not make choices? But, if so, can we afford to simply choose those texts which are familiar and nearby? The canon has exploded from too large a volume of texts.

2. There is another explosion as well that affects the canon. For today, our very histories have been challenged, often primarily in nontextual domains.

If the scientific revolutions of the last few centuries have shrunk the globe to its instantaneous form in terms of communication and travel, they have also expanded it in terms of temporal depth and complexity in other respects. Let us look at a couple of current heated debates related to our ancient anthropological histories.

A. When and where did homo sapiens, "modern man," arise? There are several theories today, one of which holds that homo sapiens may have evolved at several places as a natural evolutionary development from previous humanoids (which we know go back at least a million and a half years). The other holds for a specific appearance about 100,000 to 200,000 years ago in Africa. It now appears that the latter view is gaining precedence due to the genetic pool retro-projections and the discovery of an ancient site of sapiens in Israel contemporaneous with Neanderthal remains. But in either case, the 100,000-plus history is one undreamed of two centuries ago.

B. Or, more locally, when and how did humans appear on the American continents? Here, the debate focuses on either an arrival over the Bering Straits no earlier than 12,000 years ago, or possibly much earlier arrivals, as much as 20,000-plus years ago. (As an aside, the longest continuous culture of humans known today, ironically, is now that of the Australian Aboriginals, whose history of artifacts is recorded as far back as 40,000 years ago in southern Australia.) Again, even this now regarded "recent" history is much longer than imagination allowed only several centuries ago.

What is doubly interesting in this debate is that, in some sense, the arrival of humans on *our* continent makes them part of *our* history—or should make them part of our history, unless we simply affirm colonial history as the only history. And perhaps we can take some pride in the attainment of this history as well. Another finding of only recent archaeological studies has been that small microchips of obsidian flakes, once regarded as detritus, have been theorized to be small, delicate instruments used for artistic or, possibly, surgical purposes. The first find to confirm that theory, interestingly, turned up on a site in the American Northwest only last year. A microflake in a site dated about 11,000 years ago was preserved still attached to a stick handle. (Incidentally, that same year a find of an 11,000-year-old site near London turned up with evidence of cannibalism.)

So, I ask, which history do you want? The ancient European one complete with cannibalism, or the apparently sophisticated culture of Northwest Indians using microflakes for a detailed cultural activity? Which is *our* history?

C. One other example of comparative nature: where did the great civilizations arise, and which should be considered for the highest attainments? The standard view frequently traces our history back to the Egyptians, roughly 4000 B.C., with monumental building projects, about 2500 B.C., or the Mesopotamians of roughly the same magnitude of pastness. Yet we know today that there were vast cultures in South America with complicated irrigation systems on the western edges of the Andes, which rivaled in attainments those great civilizations. (In South America, the rising mountains, because of rapid tectonic plate activity, eventually destroyed the systems by reversing water flow.) The later and more familiar South American cultures, still operational at the time of the first conquests of the Spaniards, had systems of writing, an astronomy which regulated everything from agriculture to wars, and an empire stretching from southern North America to the depths of South America. Again, which is *our* history?

I suggest that the twenty-first century answer, perhaps a postmodern one, must be *both*. But if that is so, the canon has exploded completely in a multicultural fireworks display.

D. The above applies to "American" history. Knowledge of where we came from (Europe) needs to be supplemented by the knowledge of the place to which we came (America). But the postmodern needs even further expansion if we are to appreciate its true impact. Too often our histories have been *deliberately* insular. A striking set of examples of that insularity applies to (1) the minority traditions that lived throughout our histories in the midst of our dominant histories—I speak particularly of Jewish history prior to the twentieth century; and (2) those histories that did as much to shape our histories as those internal to us—here I speak particularly of Islamic history which, externally and then internally, shaped our Western history.

While the twentieth century thrust Jewish history upon us—in the tragic and inhumane form of the Holocaust—its antecedents are often simply subsumed or forgotten as a minority variant amidst us. Only at the points of contact or of opposition are we acutely aware of this history. In drama we know well of the stereotype of the "Jew" as Shylock in Shakespeare. And, in philosophy, we sense the strong difference of Spinoza from his early modern peers in precisely his lack of mechanism and his sense of the interconnection of all things (often thought of as a pantheist when, in fact, his roots are deeply Jewish mysticism). How little we know of life in the ghetto and the shtetl. Or, how many of you know that the school of Henry the Navigator, the school which taught the means for the voyages of discovery, was staffed by Jewish mathematicians and geographers?

Even worse, the very "enemies" who had to be repulsed—the Moors, the Ottomans, Suleyman the Great—are hidden as ghost presences motivating the Crusades. Yet through this unknown culture came (1) the revival of Greek science through the Arabic sources that revived Aristotle; (2) a wide culinary tradition which stimulated a hunger for spices which, in turn, helped motivate the voyages of discovery that eventually led to the discovery of the New World; and (3) arts, which today continue to inhabit our homes and, earlier, our cathedrals, without histories—for example, Oriental rugs, a Muslim art form par excellance.

I am, of course, suggesting that core curricula are not only cultural programs but *political* programs. They are both a selection of what is to be remembered and rehearsed, and a selection of what is to be forgotten and obliterated. But I am also suggesting that such a program runs counter to the deeper realities of the twenty-first century's emergence. Helping us change our sensibilities in this direction, particularly in the humanities, has been the spectrum of concerns from minoritarian groups: feminism, racial minorities, multiple ethnic traditions, etc.

These groups have made us aware of the occultation which has occurred within our own histories as well. In my own reading, I have become particularly interested in what happens within academic feminism. There is something like a progression to its recent, but growing, maturity. First there is an awareness of inequity and prejudice—for example, a salary differential between male and female faculty, and a pattern of nonpromotional advancement. Then, once a history is begun, "lost pasts" are discovered—why was "Rosie," who did much of the actual lab and data work for DNA, not awarded a share of the Nobel Prize when Watson got his? Or, why was not the actual first person to the North Pole—not Perry, but Matthew Henson, a Black—not given credit for the attainment? Thirdly, a growing sophistication is then attained about how perspectives are constituted, with challenges about the very ways in which things are to be seen—this latter is the *theoretical moment* of such minority histories.

These histories, so dynamically present in the academy today, indirectly also help us become aware that all histories are not only revealing histories, but also concealing histories. And the theoretical awareness of the end of this century is one generalized on by many postmodernist tendencies, which see that there is a perspectivalism involved with any single theory. What is needed is a multiple, a pluralistic, refractory set of approaches, which uncover the complexity and multiplicity of any given tradition.

Here we begin to sense something of the Other to the centripetal forces of the core curriculum as posed by Bennett and Bloom. This is the movement within the humanities today toward '"theory" and a certain interdisciplinary ferment which threatens to erode long extant disciplinary boundaries. At present, it is a centrifugal force, moving in a direction away from any canon at all.

One cannot say that it has only one face, for what today passes for "theory" in humanities circles is not only not uniform, it is also not always new in origin. For example, much theory and criticism remains bound to nineteenth-century origins—I speak here of the widely used neo-Marxian and neo-Freudian theories still much in vogue among humanist critics and interpreters. But at least these theories have the advantage of relating our histories to the humanly concrete tensions of economic ulterior motives and social-political forces which so much classical interpretive scholarship ignores or abstracts in favor of Platonic forms. Or, these theories relate our cultural myths to that life of desire and human psychological growth which repeats itself from Oedipus to *Dr. Strangelove*.

And other theories have even older roots—for example, twentieth-century hermeneutics. Here, enriched by phenomenology and existentialism—early twentieth-century movements—contemporary hermeneutics incorporates wider notions of a text, of writing, than has been seen since the medieval period itself. Or even deconstruction, a current fad among literati, draws from the sources of linguistic structuralism, itself a theory which renewed every human science from anthropology to linguistics at the middle of our century.

Nor are all these theories compatible. Quite the contrary, it is the war of theories, or, as Paul Ricoeur called it over a decade ago, the "conflict of interpretations," which characterizes contemporary humanities theory and criticism. Yet there are underlying commonalities, commonalities which I have termed postmodern in a limited sense, which do belong to all these diverse perspectives on reading and interpretation. It is to these that I shall now turn.

V. Theory in the Humanities

A clear way of conceiving what unites the various strands of late twentieth-century theory is to state the case negatively, a via negativa for defining a center.

1. Postmodern theories are usually negative with respect to a fixed canon. For some of the reasons stated above—the canon is simply too large to be inclusive in any one curriculum, the tendency is to suppress minoritarian traditions, and the cross-cultural world is too diverse to arbitrarily choose one linear development—canons are frequently criticized.

But, if the tendency to read postmodernist theory in this critical form is dominant, what could be missed is the potentially positive underside to the above critique. And that is the assumption, sometimes implicit, but often explicit, that *any text whatsoever* is open to a critical, theoretical, interpretive reading. That is, there is a kind of universality to be found within the particularity of a perspective.

2. The second critical tendency of much postmodernist theory is that which aims at eroding traditional boundaries between disciplines. Indeed, today's theorists are precisely those thinkers who are not easily defined in terms of the traditional bifurcations between philosophers, literary critics, and intellectual historians.

To be sure, this sometimes causes problems for the thinkers themselves, who are sometimes read out of disciplines too narrowly self-defined. Thus, a Richard Rorty is sometimes thought of as having left philosophy per se, a discipline that he has described as having come to an end.

Yet, here, too, there is a positive side, and that is the very interdisciplinarity with which such thinkers are associated. Rorty is today as much read in English and Comparative Literature Departments as in Philosophy. His message has become a wider, rather than a specifically disciplinary, one. Or, to put the critique positively, postmodernist theories are interdisciplinary in character, often stimulating new paradigms of thought across extant disciplines.

3. The third negative characterization of much contemporary theory is its tendency to reject "centrisms," whether by that we mean Derrida's suggestive "phallocentrism," the various minoritarian traditions which reject forms of ethnocentrism, or feminism, which rejects androcentrism.

This negative characterization, of course, supplements the attack on canons and disciplines which each have their traditions of a centrism. But what would be the positive inversion of this negative form? The answer is a *polycentrism,* or, better put, the positive evaluation of cross-cultural multiplicities. Postmodernist theorists tend to celebrate such refractive features of the late twentieth century—sometimes romantically, to be sure—in their fascination with third-world cultures, minoritarian movements, or occulted histories.

Such, then, are some of the centrifugal forces associated with postmodernist theories, which are so vocally ensconced in today's universities. And this centrifugality moves in a direction quite counter to the centripetal forces that return us to a single heritage. Thus, for every Bennett and Bloom there is also a Rorty, a Derrida, or a Jameson. And for the humanities, it is a lively, if confused, time of much debate. And, in a sense, this debate is a debate about a canon, but a radical debate about whether there is, or can be, a canon for the twenty-first century. So, once again, I return to the question of the canon.

VI. Science and the Canon

I do so by returning to the larger context from which I began, a world dominated by a science-technology culture. And I can pose the question by asking: does, or can, *science* have a canon?

When it comes to the question of the *history* of science, the answer could be positive. Science, at least our Western science, seems to have a linear history which at least metaphorically implies a canon. Arising from the prescientific speculations of the Greeks into the first theoretical natural philosophies of the Presocratics, culminating in early atomic theory

(Democritus), then through a subterranean set of routes—including a detour via the Arabs while suppressed in the Dark Ages—and reemerging in the humanist Renaissance, gaining momentum in the time of Galileo, rising to ascendancy in the enlightenment, reaching a peak of natural universalizability with Newton, undergoing a revolution with Einstein, etc., we reach the present. Such is the briefest outline of a "canonical" history of science in which there is a kind of evolution—or, if we follow the "new" philosophy of science since Kuhn—a series of revolutions, which make science what it is today.

But if the notion of a canon applies to the *history* of science, it is not at all clear that the notion of a canon applies to the *praxis* of science. For the history of science is marked by texts, by leading figures, by historical movements and countermovements, and debates between individuals and schools. Yet the self-conception of science does not match this historically commensensical notion of a possible canon at all.

1. First, it does not match a historical canon, because the leading values of science as a praxis are futural and contemporary over any past. Futural knowledge (here known as theory in its specifically scientific sense) is often even more highly valued than the current state of knowledge. But the current state of knowledge is also clearly more highly valued than past knowledge. It is this sense in which science is *radically* theoretical. Aristotelian notions of motion, phlogiston, circular orbits, while perhaps genuine achievements of theory in their historical contexts, are, for present science, merely historical oddities. They are to be judged by present standards and knowledge which, it is assumed, are better, more comprehensive, and superior to any past science per se.

Such a set of epistemological values is clearly virtually directly inverse of the Bennettian universe of canonical knowledge where it is often at least implicitly assumed that Plato was superior to or, at worst, at least equal to *any contemporary humanist thinker.* Indeed, it is heresy in the canonical sense to claim otherwise.

2. Knowledge in science constantly undergoes ongoing criticism, criticism such that any extant theory may at any time be overthrown or replaced or at least revised. This value is complementary to the first which values futural or contemporary knowledge over past knowledge. And, if the new philosophy of science (Kuhn and post-Kuhnians) is correct, often what determines a new paradigm may be a radical shift in perspective, even an inversion of what was previously taken as a stability. (For example, current tectonic theory about continent movements overthrew long-held stability beliefs about solid and immobile land in contrast to fluid and instable oceans.)

Were the same criteria to apply to canons, the result would clearly be acidic to any single canon or any unmodified canon. All canons would have to be open to revision, revolution, or replacement. Again, to apply such a practice to the humanities would be to question the canon from a different basis than previously noted.

3. And, we have seen that at least one outcome of scientific and technological history is precisely to have become acidic to what I have called *traditional* cultures. For if "primitive" cultures are, by the very definition of local territories, isolation or insulation from external influences, etc., then, by extension, larger cultural units, insofar as they also have centered or canonical cores, may equally be susceptible to canonical decay.

4. Another feature of science as a practice is its specific tendency to abstract from particulars. By this I mean that the variant disciplines within science try to encompass as wide a field of inquiry a possible—physical motion, whether of anteaters or of stars or of rolling balls—so as to reach toward a kind of universalizability.

And while the connection of this abstractive tendency is not immediately obvious as an anticanonical move, it does have such a relation. What it implies is that any body whatsoever becomes the proper subject matter for a physical inquiry. And the connection I am anticipating is that, not unlike the moves of recent humanities theory, which make any *text* subject matter for criticism and interpretation, so science has long made any object-region the subject matter of its pursuits.

It is here, however, that I now reach the apogee of this speculation on the future of the humanities in the twenty-first century. But I do so precisely at this juncture, which dialectically plays the anticanonical drive of science—which remains the larger cultural context into which contemporary humanities must fit—against the internal struggle within the humanities between centripetal and a centrifugal forces. It is to a certain antinomy which emerges in this meditation that I turn in conclusion.

VII. Some Present Antinomies

We now have reached back to the present. I have shown that in its praxical dimension, science and technology operate in a noncanonical or even anticanonical way. The result, at least with respect to traditional cultures, is often acidic, causing large modifications upon those cultures, if not downright destroying those cultures.

On the other hand, within the present Euro-American humanities, the dominant movement is one which presents itself as a return to a required, core curriculum centered upon a linear, traditional culture taken from Western thought. This movement, centripetal in direction, is contested by the newer forms of theory-oriented humanities which are centrifugal in direction.

All of these debates are present in today's universities. But here we also reach a certain paradox—at least so far as I have experienced it. And that paradox is this: I find that the science faculties frequently—in fact, usually—side precisely with the Bennett/Bloom side of the controversy with respect to core curricula and humanities requirements. This choice of sides is frequently accompanied with some degree of passion.

If I am right about science as a praxis—its methods being valued toward futural, present, and the noncanonical with respect to a past, any past—then is it not odd that the practitioners would side with the traditional humanists who invert precisely those values?

There are two possible reasons for this choice. (1) On the one hand, there is a tendency for science faculties to simply be conservative with respect to the humanities and want the humanities to be repetitions of the past. In part, this relates to a sometimes condescending attitude with respect to humanities faculty—they are to be the conveyers of *past* knowledge (implied: only we scientists discover *new* knowledge). Thus the humanities play largely a celebratory, culturally repeatable, and *service* function. There may also be an implicit fear that with respect to culture, the new may be dangerous. (2) But much deeper, I suspect, is another intuition which motivates this choice. For is it not, after all, the *traditions* of Western thought which formed the condition for the birth of modern science and technology? In this respect the antinomy is that a tradition has served—and perhaps it is felt, must continue to serve—as the *motor* which drives the expansion of Western science and technology. Were *this* tradition to change, might it not imply some change in the world of science and technology? In short, there is a vested interest in keeping the humanities "traditional" if this analysis is correct.

Now invert the issue: within the humanities there has also sprung up a series of positions characterized as "theoretical," which have taken a new direction with respect to past traditions in the humanities. These positions are postmodern with respect to several variables: they are often noncanonical—any text is susceptible of a theoretical approach; they are multicultural; and while they have often focused on minoritarian issues, what is emergent is a kind of polyperspectivalism which makes them more flexible than traditional methodologies.

What has not, however, been noted about these newcomers *is that in praxis they are interestingly parallel to some of the features of noncanonical science, although directed toward what could be called "cultural objects" rather than natural objects.* For example, although implicit, theoretical humanities approaches clearly value contemporary methodologies over traditional ones; they also leave no position uncriticized or retained as foundational, and in these two respects they parallel precisely the epistemological values of science that I previously noted.

Similarly, the "abstracting" process, which makes of all cultural objects a "text," parallels the same process that science applies to the natural world. This leaves, then, a possible parallel with the impact of scientific-technological culture: will a theoretical approach in the humanities also become the dynamic, but noncanonical, approach to human cultures which science is to natural entities?

I shall not answer these questions or attempt to resolve the antinomies I have outlined here. But I will project that these issues will be the issues of the twenty-first century as humanities come to grips with a world restructured by a technological world culture. And in such a culture, ways must be found to deal with precisely those features of multiculturality, a revalued set of multiple pasts, and a new understanding of multistable humanity. Dare humanities undergo a revolution as profound as the revolutions which launched modern science several centuries ago?

Notes

Introduction

1. In 1984 I gave a series of lectures in the Research Institute of Education at the University of Goteborg, Sweden. These were later published as a monograph, *On Non-Foundational Phenomenology*, Fenomenografiska notiser 3, Publikationer från institionen för pedagogik, 1986.

Part 1. Introduction

1. Michel Foucault, *The Order of Things: An Archaeology of the Human Sciences* (New York: Vintage Books, 1973), xxiii.

Chapter 1

1. Marco Cianchi, *Leonardo da Vinci's Machines* (Florence: Becocci Editore, 1988), 17–18.
2. Ibid., 12.
3. Ibid., 15.
4. Lynn White, Jr., "Cultural Climates and Technological Advance in the Middle Ages," *Viator* 2 (1971), 125.

Chapter 2

1. Daniel J. Boorstin, *The Discoverers: A History of Man's Search to Know His World and Himself* (New York: Vintage Books, 1985), 224–31.
2. Ibid., 209–17.
3. Ibid., 4.
4. Pacific navigation techniques have been well documented and described by Thomas Gladwyn, *East is a Big Bird* (Cambridge: Harvard University Press, 1970) and David Lewis, *We, the Navigators* (Honolulu: University of Hawaii Press, 1972). My text condenses some of these findings.

5. Boorstin, *The Discoverers,* 190.
6. Ibid., 191.
7. Ibid., 199.
8. Lynn White Jr., "Cultural Climates and Technological Advance in the Middle Ages," *Viator* 2 (1971), 177ff.
9. Boorstin, *The Discoverers,* 190.
10. Don Ihde, *Technology and the Lifeworld: From Garden to Earth* (Bloomington: Indiana University Press, 1990), 66–69.
11. Ibid., 164ff.

Chapter 4

1. See Derek da Solla Price, "Notes Towards a Philosophy of Science/Technology Interaction," in *The Nature of Technological Knowledge,* ed. R. Laudan (Dordrecht: D. Reidel, 1984).
2. See my *Instrumental Realism* (Bloomington: Indiana University Press, 1991) for a more complete account of an array of philosophers on science on instrumentation.
3. In a typical science-industry occurrence, the first published pictures of atoms were arrangements of the logo "IBM."
4. Ian Hacking in his *Representing and Intervening* (Cambridge: Cambridge University Press, 1983) has an excellent discussion of the history of the microscope; see 186–209.
5. A recent history of a large instrument experiment and a discussion of the problem of instrumental artifacts may be found in Peter Galison, *How Experiments End* (Chicago: University of Chicago Press, 1987).
6. Lee W. Bailey, "Skull's Darkroom: The *Camera Obscura* and Subjectivity," in *Philosophy of Technology,* ed. P. Durbin (Dordrecht: Kluwer, 1989).
7. H. Gernsheim, *The History of Photography* (New York: Oxford University Press, 1955).

Chapter 5

1. Within this category of critics I include Ortega y Gassett, Jacques Ellul, Nicolas Berdyaev, Herbert Marcuse, Theodor Adorno, and Martin Heidegger.
2. This thesis was first developed preliminarily in my *Technology and the Lifeworld* (Bloomington: Indiana University Press, 1990).
3. Although my *Technics and Praxis* (Dordrecht: D. Reidel, 1979) developed this theme, since that time a number of well-known philosophers of science have also picked up this emphasis. For a comparative study, see my *Instrumental Realism* (Bloomington: Indiana University Press, 1991).
4. See Lee W. Bailey, "Skull's Darkroom: The *Camera Obscura* and Subjectivity," in *Philosophy of Technology,* ed. P. Durbin (Dordrecht: Kluwer, 1989).

5. This line of interpretation of Galileo is developed by Derek de Solla Price, "Notes Towards a Philosophy of the Science/Technology Interaction," in *The Nature of Technological Knowledge,* ed. R. Laudan (Dordrecht: D. Reidel, 1984).
6. Many of these materials concerning Heisenberg have recently been declassified. They were made available to me by a historian colleague, David Cassidy, and were cited in my *Technology and the Lifeworld.*
7. A succinct history of engineering technology which well documents the disjunction between science theory periods and technological innovation may be found in L. Sprague de Camp, *The Ancient Engineers* (New York: Dorset Press, 1963).
8. Lynn White, Jr., "Cultural Climates and Technological Advance in the Middle Ages, *Viator* 2 (1971). See also my "The Historical and Ontological Priority of Technology over Science," in *Existential Technics* (Albany: SUNY Press, 1983).
9. See E. Ann Kaplan, *Rocking Around the Clock* (London: Methuen, 1989).

Chapter 6

1. Jacques Derrida, *Speech and Phenomena,* trans. David Allison (Evanston: Northwestern University Press, 1973), 103.
2. Michel Foucault, *The Order of Things: An Archaeology of the Human Sciences* (New York: Vintage Books, 1973), 35.
3. Ibid.
4. Cf. Edmund Husserl, *Ideas: General Introduction to Pure Phenomenology,* trans. W. R. Boyce Gibson (London: Allen and Unwin, 1931). See especially sec. 70.
5. It is to be noted that the figure/ground relation developed by the gestaltists was, in fact, originally inspired by Husserl.
6. Maurice Merleau-Ponty, *Essential Writings,* ed. Alden Fisher (New York: Harcourt, Brace and World, 1969), 48.
7. Ibid.
8. Maurice Merleau-Ponty, *The Visible and the Invisible,* trans. Alfonso Lingis (Evanston: Northwestern University Press, 1968), 212.
9. Martin Heidegger, *Being and Time,* trans. E. Robinson and J. Macquarrie (New York: Harper and Row, 1962), 95
10. See Don Ihde, *Experimental Phenomenology* (New York: Capricorn Books, 1977).

Chapter 7

1. Michel Foucault, *The Order of Things: An Archaeology of the Human Sciences* (New York: Vintage Books, 1973), 35.
2. Ibid., 38.

3. Ibid.
4. Ibid., 40.
5. Ibid., 79.
6. Paul Ricoeur, *The Conflict of Interpretations,* ed. Don Ihde (Evanston: Northwestern University Press, 1974), 33.
7. Ibid.
8. Ibid., 41.
9. Ibid., 47.
10. Ibid., 48.
11. Paul Ricoeur, *Time and Narrative,* vol. 2, trans. K. McLaughlin and D. Pellauer (Chicago: University of Chicago Press, 1984–88), 160.
12. Ibid.
13. *Time and Narrative,* vol. 2, 53.
14. Ibid.
15. Ibid., 54.
16. Ibid., 56.
17. Ibid., 77.
18. Ibid., 78.
19. Ibid.
20. Ibid.
21. Ibid., 78–79.
22. Ibid., 80.
23. Ibid., 81.
24. Don Ihde, *Existential Technics* (Albany: SUNY Press, 1983) 163–64.

Chapter 8

1. Martin Heidegger, *Poetry, Language, Thought,* trans. Albert Hofstadter (New York: Harper and Row, 1971), 42.
2. Ibid., 42–43.
3. J. Donald Hughes, *Ecology in Ancient Civilizations* (Albuquerque: University of New Mexico Press, 1975), 1.
4. Ibid.
5. Martin Heidegger, *Basic Writings,* ed. David Farrell Krell (New York: Harper and Row, 1977), 315.
6. Martin Heidegger, *Parmenides* (Frankfurt: Klostermann, 1982), 118–19. Translation by Michael Heim in *Electric Language* (New Haven: Yale University Press, 1987).
7. Heidegger, *Basic Writings,* 297.
8. Heidegger, *Parmenides,* 118–19.
9. Heidegger, *Poetry, Language, Thought,* 152–53.
10. Ibid., 152.
11. Ibid., 152–53.
12. Ibid., 153.
13. Ibid., 158–59.

14. Heidegger, *Basic Writings,* 307.
15. Ibid., 315.

Chapter 9

1. Andy Pickering has developed the "science as knowledge"/"science as practice" distinction to mark the sixties change after Kuhn and kin. Indeed, the discussion of the "strong program" as an outcome of this shift is interestingly debated in *Social Studies of Science* 20, no. 4 (Nov. 1990). See especially A. Pickering, 682–728.
2. Steve Fuller, "The Philosophy of Science since Kuhn: Readings on the Revolution That Has Yet to Come," *Choice* (Dec. 1989), 595–601.
3. Evelyn Fox Keller, *Reflections on Gender and Science* (New Haven: Yale University Press, 1985), 3–4.
4. Ibid., 10.
5. Ibid., 11.
6. Ibid., 24.
7. Ibid., 36.
8. Ibid., 136.
9. Quoted by Sandra Harding in *The Science Question in Feminism* (Ithaca: Cornell University Press, 1987), 311.
10. Fox Keller, *Reflections on Gender and Science,* 177.
11. Harding, *The Science Question in Feminism,* 251.

Chapter 10

1. William J. Bennett, "To Reclaim a Legacy: A Report on Humanities in Higher Education," NEH report (Nov. 1984), 13.
2. Ibid., 30.
3. Ibid., 1.

Also by Don Ihde

Author

Philosophy of Technology: An Introduction
Instrumental Realism: The Interface Between Philosophy of Science and Philosophy of Technology
Technology and the Lifeworld: From Garden to Earth
Consequences of Phenomenology
Existential Technics
Technics and Praxis: A Philosophy of Technology
Experimental Phenomenology
Listening and Voice: A Phenomenology of Sound
Sense and Significance
Hermeneutic Phenomenology: The Philosophy of Paul Ricoeur

Editor

Paul Ricoeur, *The Conflict of Interpretations*

Co-editor

Descriptions (with Hugh Silverman)
Hermeneutics and Deconstruction (with Hugh Silverman)
Interdisciplinary Phenomenology (with Richard Zaner)
Dialogues in Phenomenology (with Richard Zaner)
Phenomenology and Existentialism (with Richard Zaner)